Disconnected Rivers

Disconnected Rivers

Linking Rivers to Landscapes

ELLEN E. WOHL

YALE UNIVERSITY PRESS NEW HAVEN & LONDON

Designed by James J. Johnson and set in E+F Scala type by Tseng Information Systems, Inc.
Printed in the United States of America.

Library of Congress Cataloging-in-Publication Data

Wohl, Ellen E., 1962–
Disconnected rivers : linking rivers to landscapes / Ellen Wohl
p. cm.
Includes bibliographical references and index.
ISBN 0-300-10332-8 (cloth : alk. paper)
1. Stream ecology—United States. 2. Watershed management—United States. 3. Stream conservation—United States. I. Title.
QH104.W637 2004
577.6′4′0973—dc22 2004011610

A catalogue record for this book is available from the British Library.

The paper in this book meets the guidelines for permanence and durability of the Committee on Production Guidelines for Book Longevity of the Council on Library Resources.

10 9 8 7 6 5 4 3 2 1

For M, who brought me
to connections beyond science

Contents

Preface

I grew up among rivers. When I was a child, a small creek flowed behind our house in Ohio. Deeply entrenched in a ditch, its quiet waters remained thickly green with algae through the summer. I was forbidden to play in the creek water. Why, I didn't know; somehow the water was unclean.

A half hour's walk from our house lay the Rocky River. True to its name, the river wound between steep cliffs of shale and sandstone and flowed across long, flat bedding planes of rock that were always slick with algae and river slime. Here, too, I was not allowed to enter the river without wearing protective rubber boots. I watched other children swim in the river pools during hot summer days, spouting river water from their mouths, and I wondered whether they would be poisoned.

My parents bought and read Rachel Carson's *Silent Spring* when it came out in 1962, the year of my birth. My father, an environmentalist before the word became widely used, taught biology and chemistry. He knew that during hard rains domestic septic systems commonly overflowed, sending raw sewage into the Rocky River. He also knew what substances industries dumped into the waters and the air. He grew up in Cleveland, where his father worked as a machinist in a tool and dye factory and his mother worked as a seamstress in a garment factory. Those factories were located in the industrial heart of the city along the Cuyahoga River called the Flats. The Flats seemed to me as a child to be a landscape of death with its harsh, angular lines and overlay of black. It was here that the Cuyahoga River caught on fire when I was seven years old, and my home became a source of national embarrassment. Other cities joked that Cleveland was the "Mistake on the Lake."

When he began teaching in the Cleveland public school system, my father devised experiments for his students. For one, he placed the body of a vacuum cleaner in a wooden box fitted with a nozzle. Inside the nozzle he placed a square of gauze. For an hour each day the students ran the machine outside, with its nozzle pointed at the sky, and particulate matter sucked in with the air was left on the gauze. After placing the gauze between two strips of glass, the students positioned a filmstrip projector at a precise distance and shined its light through the gauze, measuring the light's intensity with a light meter at an equally precise distance on the other side of the gauze. Those squares of gauze with their unmistakable dark centers impressed me tremendously. On a good day, the gauze might have a barely discernible circular brown smudge. On a bad day, the circle made from the air near this school surrounded by aircraft and automobile manufacturers could be a vivid orange or a dense black. My father's invention won him a teaching award, and his students did well at regional science fairs.

My father also measured water quality. In some of my earliest science projects with him we measured pH, temperature, and dissolved oxygen and inventoried the species of insects, amphibians, and fish present in the rivers and the vernal ponds. He belonged to a naturalist club and taught me to identify the plants and animals of the eastern deciduous forest. My environmental consciousness took shape during our Sunday morning walks along the Rocky River; one spring I was outraged to see large mounds of disturbed sediment along the river. I assumed the mounds were the work of bulldozers. It was only much later that I realized that the spring floods had been recontouring the riverbanks.

Inexorably, I came to realize that human activities had impoverished the river worlds all around me. Lake Erie was another place I was forbidden to swim, and I could not eat the fish that came from there. It shocked me to read Edwin Way Teale's description in *Journey Into Summer* of the "mayfly storm" on the shores of Lake Erie. Stopping in the town of Sandusky in the late 1950s, Teale described the annual early summer emergence of the aquatic insects. Clouds of insects rising like smoke from the evening waters of the lake moved inland. The weight of thousands of mayflies bent down the leaves of plants. Spent mayfly bodies collected along the waterline in windrows three feet deep, and traffic slowed down on streets slippery with crushed mayflies. There were so many million and billions of mayflies that "numbers lost their meaning." I had never heard of such an occurrence. My parents confirmed that such phenomena used to occur, before the bottom sediments of the lake became so polluted.

Shortly before we moved from Ohio, the second-growth woodlands behind our house were cleared for development. A church replaced the stands of fast-growing maple and oak, and a shopping center obliterated the cattail marsh where the mallards nested. The little creek was directed into a drainage pipe.

We moved to central Arizona, where the rivers now flow only after rain. I attended the university to study geology. I read of the early European American settlers along the Salt River, who used the river's flow to run a flour mill. Malarial mosquitoes had bred in the wetlands along the river. I wandered out from my dormitory for walks along the dry bed of the Salt, observing the skeletal anatomy of the river and growing indignant over environmental change.

I graduated from the university and moved to Colorado, where the rivers still flow. I wandered delightedly up the canyon of the Cache la Poudre River, thinking of it as the pristine mountain river where French fur trappers cached their gunpowder in the 1820s. Then I read early accounts of the river and realized that I could not find the beavers, the lush riverside forests, and the abundant trout that early settlers had described. I wrote my own book, *Virtual Rivers,* to prevent others from mistakenly assuming that the rivers of the Colorado Front Range are pristine.

Along with rivers, I also grew up among books. Rivers I could experience directly gave me only partial insight. I needed the broader and deeper perspective of books to understand how rivers function, how those functions have changed through time, and our role in those changes. As I walk beside a clear-flowing mountain creek, I can see through the water to the texture of the streambed. I can see cobbles green with algae, and tan-colored sand beside the cobbles. A fish darts above the cobbles, creating a swift, dark shadow. Mossy banks and rotting logs overhang the creek. I cannot see the aquatic insects clinging to the cobbles, their delicate nets spread against the current to catch microscopic animals drifting downstream. I cannot see the calcium that dissolved in the water where it flows underground before emerging upstream as a spring. I do not notice the grassy ridge of boulders left when a debris flow completely recontoured the creek thirty years ago, let alone the bedrock knobs smoothed by a glacier during the last glaciation. My direct perceptions are limited to the most obvious features, to the human scale of detail, and to the moment when I observe the creek. And even in that moment, at my own scale, I perceive what my experience has prepared me for; I perceive the arrangement of cobbles and sand on the streambed because I study these phenomena, but I do not perceive the species and ages of the riverside trees. But when I walk the creek with a botanist, she shows me how the river birches are all the same age, having germinated after a flood swept the banks of the creek clean and left fresh sediment for seedlings. An aquatic ecologist brings his collecting net and shows me the astonishing variety of tiny, squirming creatures living out their lives under the cobbles. A historian pulls aside a tangle of saplings to reveal the crumbling foundations of a cabin built during the past century by placer miners working the bed of the creek for gold. As I draw on this knowledge, I build an understanding of the river ecosystem through time and space. Much of that understanding comes from indirect knowledge, from insight gleaned through books. We all owe a huge debt of gratitude to those who care enough to study some part of a river deeply and carefully and then to share their insight through the written word. And so I go back and forth, from river to book to river, trying to understand.

Developing this book has become a personal experience. My perceptions of the world around me changed as I wrote it. Since writing the chapter on water pollution, for instance, I have altered my diet to eat mostly organic, vegetarian foods. If I drive through an agricultural landscape, I no longer perceive benign, pastoral scenery. Instead, I feel an urge to roll up the car windows and hold my breath as I think of the poisons broadcast over the fields. I have become skeptical about the treatment of the food I eat, and of the neatly manicured lawns of my neighbors and my university campus.

Stephen Pyne traced the beginnings of a philosophy of water conservation in the United States to geologists such as John Wesley Powell. Powell's vision of water use was grounded in sustainability and the limits of the understanding of nineteenth-century science. As those scientific limits expanded during the twentieth century, we came to understand more quantitatively what John Muir and others intuited long ago—that everything is interconnected, and sustainable human societies cannot exist apart from sustainable ecosystems.

We all live among rivers. They are the sinews that bind our landscapes together. I have come to feel with increasing urgency that as we unwittingly strain or cut those sinews, we threaten the integrity of the whole environment on which we depend. I grew up in a wet countryside where it was difficult to clean the water enough to drink it. I now live in a dry countryside where it is rapidly becoming difficult to find enough water to drink. We have taken rivers for granted for centuries, and we continue to do so at our peril. I do not think we can continue in this manner for much longer. I hope that we do not try to. This book is an expression of that hope.

Acknowledgments

Writing this book has provided me with both pleasure and an education. The pleasure has come in part from the opportunity to discuss rivers with others who are passionate about rivers, and about writing. Hiroshi Ikeda of Tsukuba University invited me in 1997 to give a talk, "Americans and Rivers," at an international symposium held on the Kurobe River in Japan. That presentation formed a nucleus that grew into this book when Jean Thomson Black of Yale University Press suggested that I write about rivers. Jean has continued to work with me during all stages of crafting a sketchy idea into a finished book, and I much appreciate her friendship and skill. Several people read drafts of individual chapters: Madeleine Lecocq read the introductory and concluding chapters; L. Allan James read the chapter on pioneer impacts; Brian Bledsoe read the chapter on commercial impacts; Douglas Smith read the chapter on bureaucratic impacts; and Douglas Thompson read the chapter on river restoration. Each of them provided insightful and detailed comments that substantially improved the book. Douglas Thompson, Judy Meyer, and two anonymous reviewers also reviewed the entire book at the request of Yale University Press. Their careful reading and analysis further integrated and strengthened the book. The staff of various government agencies and library collections were very helpful in collecting historical photographs, and I thank those at the Bancroft Library, University of California at Berkeley; the Cleveland Public Library; the Florida Department of State, Division of Library and Information Service; and the South Florida Water Management District. Joe Tomellerie graciously provided the drawing of a paddlefish, Peter LaTourrette provided the photograph of a dipper, and the Missouri Department of Conservation provided the photograph of a hellbender. My parents, Richard and Annette Wohl, were with me each step of the way, telling stories of growing up in Cleveland and gently prompting me with their favorite question, "When is your book going to come out?" Thank you all.

Disconnected Rivers

Chapter 1

Why Should We Care About Rivers?

Rivers reflect a continent's history. Where forces far beneath the Earth's crust force up mountain ranges, rivers flow swift and cold down steep, boulder-strewn channels. Where the Earth is still, rivers meander broadly, depositing thick plains of sand, silt, and clay.

They also reflect a people's history. Where people clear the forests for agriculture, river valleys retain sediments, recording the transitional period when the soil washes down from the hillslopes, and rivers become broad and shallow. Where people mine precious metals from hills or build electronics factories, river valley sediments contain the toxic by-products of these activities. People build canals, roads, and railroads along river corridors, following river passages through dense forests or steep mountains.

River valley sediments record all the changes in a river's drainage basin over thousands of years. The river itself records the most recent changes, steepening its course as it crosses the furthest sediments deposited by a glacier now melted, or dammed where farmers in the 1950s wanted water storage.

The organisms living in and along rivers also reflect history. Along a river downstream from a site where mining occurred in the 1890s, there are fewer individuals and species of aquatic insects and fish in the twenty-first century because toxic metals still leach from the mining site. Where a river repeatedly shifted its course back and forth across the valley bottom during floods spread across 200 years, cottonwood seedlings have sprung up on each new sandbar created by a flood. Now the river has groves of cottonwoods aged 10, 40, 80, and 175 years, and these trees map the changes in the river's course. Where dammed water released from the bottom of a reservoir creates a cold, clear flow, introduced trout thrive, out-competing the native fish adapted to the warm, sediment-laden waters present before the dam was built. And where a dam blocks native fish returning from the sea to spawn, these fish are no longer present at the headwaters of the river.

The physical forms of rivers and river ecosystems are our historical archives, yet these archives are challenging to interpret. Gaps may be present in the physical record where sediments deposited during an earlier period of river history were subsequently eroded. Because of the gaps we can seldom decipher a complete and continuous record of a river's history. But by assembling the records from many rivers we can piece together regional and continental syntheses of history. Organisms living in and along a river also have an evolutionary history, and the unique evolutionary lineages present in different river drainages provide clues to the history of isolation or integration of each drainage.

We owe rivers the respect due to any source of information that helps us to understand our history, and so to understand ourselves. But rivers are also our lifelines. They provide us with the water we drink, the water that helps our crops to grow, and the water that fuels or cools our industries. Water is a universal solvent and is used at some stage in the manufacture of every product we consume. Rivers transport our wastes, and to some extent transform them. If not for this self-purifying function of rivers, many of our estuaries and deltas would be even more polluted. Rivers transport our goods, generate our power, and sustain our recreation. The condition of our rivers, more than any other natural resource, reflects our attitudes toward the world around us, and ultimately our attitudes toward ourselves. The society that does not protect its rivers destroys its own lifelines.

This book draws a connection between lack of respect for rivers and lack of understanding of rivers. We in the United States have not fully appreciated the vital functions that rivers perform.

Human beings have used the natural systems of America as resources since the first people reached this continent. Such use reaches unsustainable levels whenever people deplete a physical resource to the point that the resource is effectively no longer available to them, or whenever they deplete a biological resource to the point that a species can no longer sustain itself or perform its ecological functions. Unsustainable irrigation practices led to the salinization of Hohokam agricultural fields in the southwestern United States by A.D. 1300. Unsustainable hunting, trapping, and fishing by European Americans led to the extinction or near-extinction of beaver in the eastern United States during the 1600s, the bison by 1880, the passenger pigeon in 1914, and commercial fisheries for shad, cod, sturgeon, and other species during the later nineteenth and twentieth centuries. Unsustainable logging practices altered forest ecology and composition in a manner that will require centuries to overcome in regions as diverse as the southern Appalachians (1900s–1930s), the Colorado Rockies (1860s–1890s), and the California coast ranges (1950s). Unsustainable river flow regulation and diversion led to the endangerment of native fish, massive loss of riverside vegetation, erosion of archeological sites, and degraded water quality along the Colorado River as well as to the collapse of salmon populations in the Pacific Northwest.

This long history of resource destruction is partially offset by a developing vision of resource conservation. The environmental movement in the United States has emphasized the conservation of natural resources and the reservation of public lands since the late nineteenth century, when leaders such as John Muir, Gifford Pinchot, and Theodore Roosevelt persuaded the federal government to designate the first forest reserves and national parks. Public support for conservation grew during the succeeding century, along with concerns about how human activities might be affecting the world around us. Development of the modern conceptual framework of ecology during the 1960s and 1970s emphasized that the environmental health of public lands must be assessed in terms of the physical habitats that support communities of interdependent species. With increasing concern over the number of species becoming endangered or extinct, Americans are realizing how closely the species present in any community are linked to the physical landscape and to one another by numerous chemical and physical exchanges. We cannot save an endangered species of trout without also saving the river and floodplain habitat in which that trout evolved, as well as the plants and insects that form the food web in which that trout exists.

At the same time that the federal government reserved public lands, it also invested in land reclamation and engineering, building dams to irrigate agricultural lands and control floods and dredging and straightening rivers that were then confined within levees. While the government created legislation guaranteeing the quality of air and water and protecting endangered species, it also subsidized road building, timber harvesting, mining, and grazing on public lands.

The United States thus has two competing traditions. One tradition emphasizes individual and corporate freedom to optimize short-term profits, with economic growth and increased standards of living based on the excessive exploitation of natural resources. The other tradition emphasizes resource conservation and environmental protection, an interest in natural history, and expectations of outdoor recreation and public access to wilderness areas. These competing traditions together shape the understanding and use of rivers in the United States.

Despite the history of public awareness of environmental issues in this country, many people remain unaware of how substantially human activities have altered rivers across the nation. Human activities affect the movement of both water and sediment along a river—the river's form. Human activities also alter the river's ability to provide habitat and nutrients for diverse species—the river's function. The distinction between form and function is important because it governs public perceptions of rivers.

The form or physical appearance of a river can be readily perceived. People commonly expect a "healthy" river to be "pretty"—to have clear water, stable banks and bed, and perhaps a fringe of trees along its banks or fish in its pools. These expectations of a healthy river's appearance may be misleading in that they ignore loss of function. However, it is difficult to assess a river's function with only a casual examination. River channels are fundamentally conduits for water and sediment, but the specific processes of water and sediment movement vary widely among channels. These processes create unique habitats and processes of nutrient exchange to which the local in-channel and floodplain communities of plants and animals are adapted.

The channel bed of a natural river, for example, is unlikely to be of uniform depth or material for more than a few tens of yards downstream. Most channels have alternating deep pools and shallow riffles. The riffles have coarser sediment and faster and shallower water. Species of aquatic plants, insects, and fish adapted to rapid, shallow water favor riffles, whereas a few yards upstream a different community of species will inhabit the deeper waters of a pool. Because these differences are not readily apparent to an observer, a river with severely impaired function may appear to be healthy. A river with clear water and stable banks

View upstream along the Savage River in Denali National Park, Alaska. This river is fully connected to the surrounding landscape and richly diverse in form and function. Pools and riffles alternate downstream, and riverside vegetation forms a mosaic of different types and ages. During floods, water flows across the floodplain, submerging surfaces such as the cobble bar in the right foreground.

supporting a few mature cottonwood trees meets many people's expectations of a picturesque river. But clearing of wood from the river channel may have destroyed the pools and riffles, changing a diverse in-channel habitat supporting numerous species to a largely uniform channel supporting only a few species. Flow regulation associated with upstream dams may have altered the river's flow such that the banks are more stable and cottonwoods, which germinate on freshly deposited sandbars, are no longer reproducing. Changes in river form have led to an impoverishment of river function and a decrease in biological diversity. If we do not understand how a natural river would really appear or how it would function, however, we will not recognize when a river environment has been altered.

We cannot save trout without saving their river and floodplain habitats. We cannot save river and floodplain habitats—and the plants and insects of the trout's food web—if we do not also maintain the processes controlling water and sediment entering the river corridor from

View upstream along a small river with placer mining in central Alaska. This river has been heavily impacted by increased sediment and reduced streambed and bank stability. Pools and riffles have been buried under sediment, and riverside vegetation has been destroyed. Both form and function have been impoverished.

the surrounding hillslopes and uplands. They go hand in hand. A functional river ecosystem is connected to everything around it: the atmospheric and oceanic circulation patterns that control precipitation over the drainage basin; the soils developed on the hillslopes adjacent to the river during thousands of years of weathering of the underlying bedrock; the plant communities growing on those soils, and the animals that pollinate and consume the plants; the processes by which precipitation filters down to the groundwater and raises or lowers the water table that is intimately connected to most streams; and on and on. By altering our river systems we have, in many cases, severed these vital connections. Dams interrupt the upstream-downstream passage of fish, the downstream flow of seeds that replenish riverside forests, and the downstream movement of water and sediment. Timber harvests short-circuit the gradual downslope flow of rainwater below the ground, instead sending masses of water and sediment quickly into nearby rivers. Artificial levees keep young fish from the rich nursery habitats created by warm, shallow waters spreading across a floodplain during high flows and prevent the pulse of nutrients returned to the channel as floodwaters recede. Disconnected rivers become impoverished in form and function because the processes maintaining form and function no longer operate. We increasingly have discovered that it is enormously expensive and difficult to artificially re-create these processes—to pass salmon through a dam with fish ladders, for example, or to stabilize and revegetate clearcut hillslopes. River corridors function most fully and effectively when they remain connected to the total environment.

Today, the American people are being asked to make decisions regarding rivers from the national level (in relation to the Clean Water Act, the Wild and Scenic Rivers Act, the National Floodplain Insurance Act, the Endangered Species Act, and other federal legislation) to the local level (in relation to wastewater management, nonpoint source pollution, in-stream flow, flood hazards and community zoning, open space and greenbelts, recreation, and dam relicensing and removal). Most people have an instinctive appreciation for flowing water and look to rivers as a source of recreational and esthetic enjoyment. But others still regard rivers as mere conduits for the transport of commodities and wastes or as natural hazards that must be controlled. These conflicting ways of seeing rivers and hence demands on river resources only intensify as both global population and U.S. population and resource use continue to grow. If we are to make informed decisions regarding rivers, it is important that we learn to think of rivers in terms of both form and function. This requires that we look beyond the obvious physical characteristics of a river and think of it as an extensive ecosystem interdependent with and connected to the surrounding floodplains and drainage basin.

Chapter 2

American Rivers

The rivers of the United States are as diverse as the country's people. The rivers meander slowly across marshy plains hazy with heat and humidity. They rush down steep, rocky gorges fed by the melting ice of glaciers. They flow hidden beneath the ground in limestone caves, or they flow only after a thunderstorm has abruptly saturated the desert's surface.

Animals adjust themselves to this diversity. Silvery salmon swim relentlessly up clear, cold waters to lay their eggs among gravels eroded from the jagged Sawtooth Mountains of Idaho. Catfish with barbels sensitive to subtle movements in murky water wait beneath overhanging banks of clay in the lowlands of Louisiana. Mallard ducks call to one another as they land on the gray-green waters of the upper Mississippi. Beavers work steadily with slender branches and saplings, anchoring them among cobbles pushed up into a ridge, and overnight a small mountain stream in Utah is dammed.

Starting at least twelve thousand years ago, humans also adjusted to this rich diversity of river environments and shaped the rivers to their own needs. Human impacts to rivers depend in part on the characteristics of each river ecosystem, and these characteristics reflect the geology, climate, and history of a river basin.

The Physical River System

A river channel conveys water and sediment downstream. Imagine two hypothetical rivers: one flowing from the Rocky Mountains in the southwestern United States, the other flowing from the southern Appala-

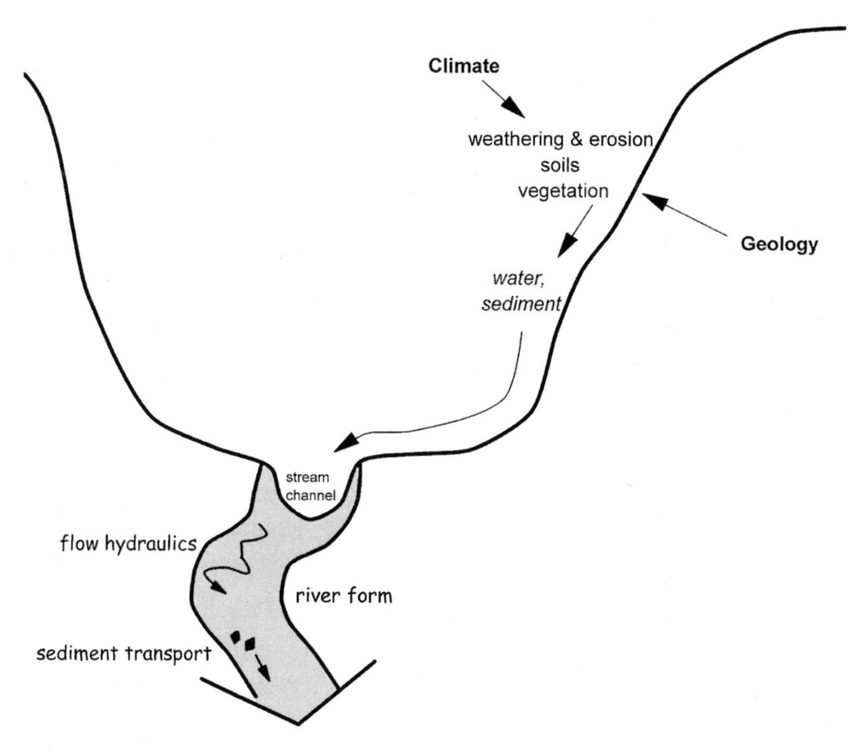

The physical river system. Geology and climate together determine how bedrock is broken down into sediment that creates soils and supports vegetation. Water and sediment moving downslope into a river govern the form of the river. River form in turn influences the manner in which water and sediment move downstream.

chians. The contrasts between these rivers illustrate how geology and climate interact to govern the amount of water and sediment supplied to a river, and how the water and sediment supply control river form and process. Both rivers begin in steep terrain where a long history of uplift contorted the bedrock into mountains.

Faulting and the intrusion of large masses of molten rock into the overlying crust during millions of years created the Rockies. Alpine glaciers ground away at their upper valleys until about ten thousand years ago. As the ice widened the valleys, it also deposited huge boulders along them and, when it melted, sent much higher volumes of water and sediment down the rivers. The granite underlying the drainage basin resists weathering and erosion in the relatively dry climate present since the retreat of the glaciers. The tough rock slowly weathers to cobbles and gravel that form steep slopes below the bedrock cliffs. Where softer rock such as shale is present, weathering produces silt and clay that form rounded, gentle slopes.

Summer rains fall with such swift intensity that the rocky, unvegetated ground does not absorb the rainwater, which runs quickly downslope into nearby streams. The water carries large amounts of sediment with it because semiarid regions such as the southern Rockies have enough precipitation to at least periodically move sediment downslope but not enough vegetation to stabilize sediment on hillslopes. The rapid but episodic movement of water and sediment into the river occasionally produces flash floods or debris flows, but even these may not be strong enough to move the large boulders left in the stream channels by the melting glaciers. Floods and debris flows keep the river broad and shallow. In places the river braids into multiple channels. Flow is turbid, with sediment carried in suspension by the water, and only those plants and animals that can withstand such turbidity and rapid fluctuations in flow level live within and along the river. As the river leaves the mountains and enters the broad desert basin beyond, the flow has only enough energy to carry the smaller sediments. The boulders and cobbles of the mountain channel give way to sand and gravel farther downstream. The river may cease to flow on the surface as water percolates into the thick layer of sediment underlying the basin.

The Appalachians are older mountains, created as compression between North America and Africa folded bedrock into tight ridges and valleys. The milder climate supplies abundant rainfall, and the softer sedimentary rocks of the Appalachians weather in this warm, wet climate to thick soils. The soils support dense forests that help to lessen the impact of falling rain, allowing the rainwater to be absorbed by the soils and move downslope more slowly beneath the surface. The Appalachian river has a more constant flow than its counterpart in the Rockies and carries less sediment. The river is narrower and flows in a channel lined with dense riverside forest. Where the river crosses harder rocks, the downstream slope of the valley is steeper and the river has a stair-step configuration. Softer rocks weather to a lower downstream slope, and the river alternates downstream between riffles and pools. As the river leaves the mountains and enters the adjoining foothills, the downstream slope of the valley decreases further, the valley grows wider, and the river meanders across the valley bottom.

View upstream along a meandering river in North Park, Colorado. The river is constantly shifting back and forth across its floodplain as individual meander bends erode along the outside of the bend and fill with sediment along the inside of the bend. Occasionally, the channel will straighten its course across a bend, leaving the abandoned meander as an overflow channel or isolated pond. In this photograph, the band of darker riverside vegetation indicates the width of floodplain across which the river meanders. This width is several times greater than the width of the actual channel.

As geology, climate, and topography change with time or downstream, river form changes in response. A river is thus self-adjusting and variable over time and space. These adjustments may be temporary, as when a river widened by a large flood gradually narrows during the succeeding decade; or the adjustments may be longer-lasting, as when a housing subdivision and its accompanying pavement decrease sediment and increase water supply to a river, causing the river to permanently widen.

Because a river continually responds to changes in its environment,

A closer view of an individual meander bend along a river in the Wind River Range of Wyoming. The outside of the bend, where the flow is swifter and deeper, is eroding. The shallow inside of the bend has a bar of lighter sediment recently deposited by the river.

it is never static. The type of river response depends on the magnitude and persistence of changes in water and sediment entering the river. The movements of water and sediment within a sand-bed channel adjust readily over a period of minutes to hours. As discharge increases during a flood, for example, flow velocity increases, and the greater force of flow exerted against the streambed and banks brings sediment into transport. As the floodwaters recede, much of this sediment is redeposited along the river. If the change in water and sediment entering the river is more dramatic or longer lasting, even a river formed in bedrock can alter its form and downstream slope in response to the change in supply. This adjustability of rivers is at the heart of how rivers respond to human land uses that alter water and sediment supply to stream channels.

River Regions

The diversity of river forms present in the United States and North America reflects the patterns of geology and climate present across the

View upstream along a mountain river in northwestern Montana. This steeper channel has a stairstep configuration, with steps formed of boulders and small pools alternating downstream. Turbulent mixing in each pool keeps the water highly oxygenated. Such stairstep configuration is characteristic of headwater streams.

continent. The continental United States is drained by five major river systems—the St. Lawrence, the Mississippi, the Columbia, the Colorado, and the Rio Grande—and by numerous smaller river networks. The movement of water and sediment from the continent to the oceans is not evenly distributed among these rivers. Much of the water comes from the wetter regions, and the sediment from the drier regions. The combinations of geology and climate produce six distinctive "river regions" in the United States. River form, flow characteristics, and sediment transport are broadly similar within each region.

A pool formed as flow plunges over logs jammed across a river in Washington. Notice the differences in streambed sediment, from finer gravels at the right to boulders in the pool. Such differences create diverse habitats for aquatic insects and fish.

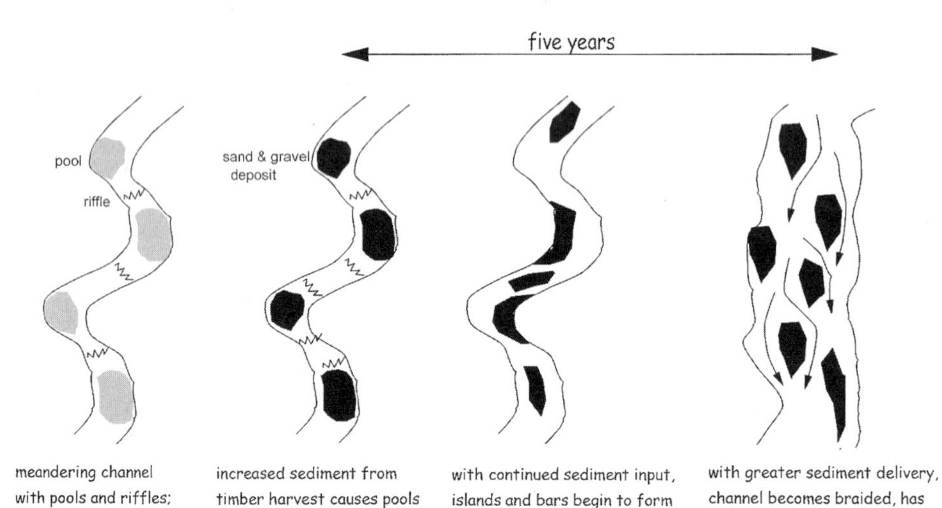

An example of river adjustment.

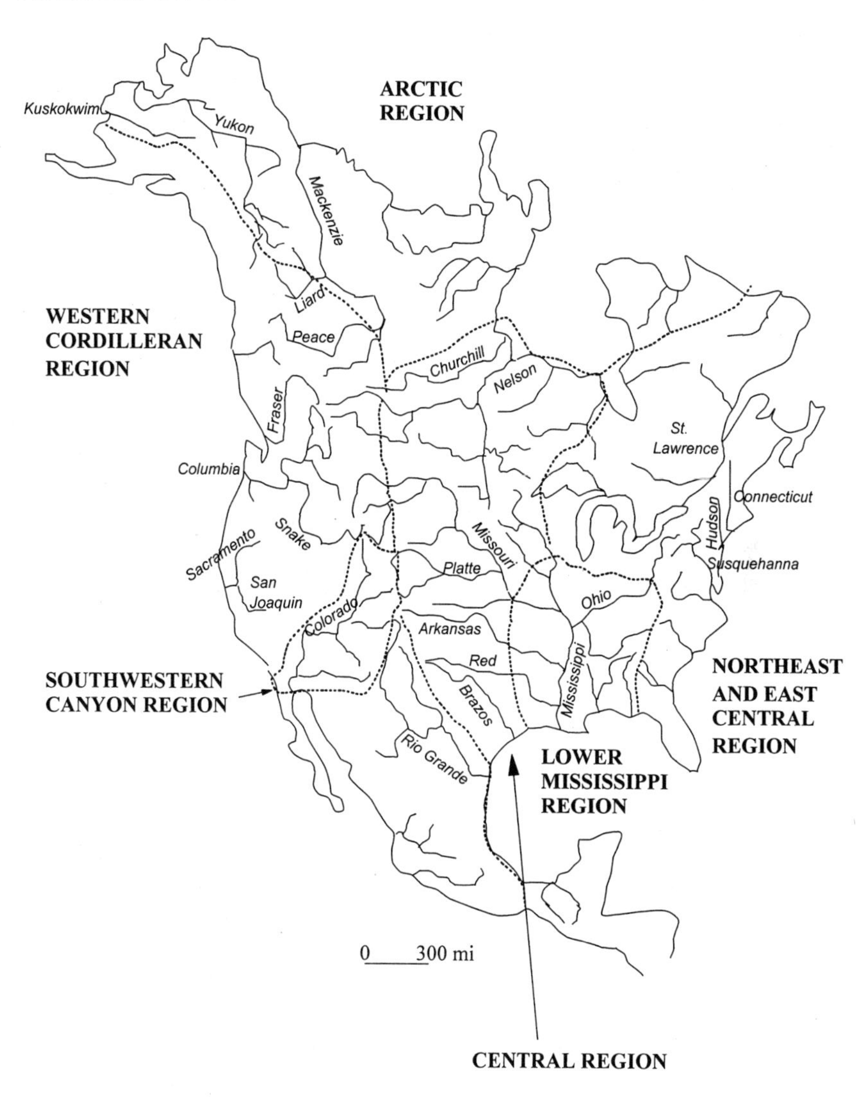

River regions and major rivers of the United States.

Rivers of the Northeast and East-Central Region

Rivers of the Northeast and East-Central region drain east to the Atlantic Ocean. The single largest drainage basin in this region is the St. Lawrence, which delivers an average of sixty-five thousand billion gallons of water to the ocean each year. South of the St. Lawrence, the rivers originate in the Appalachian Mountains and cross the Piedmont and coastal plains before reaching the ocean. The flow of these rivers peaks

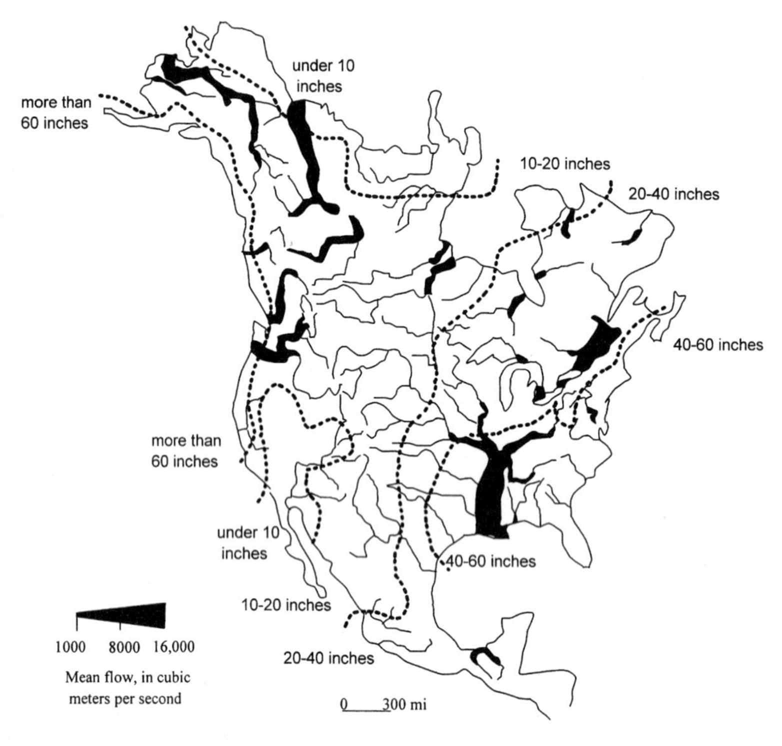

Relative contributions of total flow (left) and sediment (right) from North American rivers, with approximate average annual precipitation zones. The Mississippi, St. Lawrence, Mackenzie, Yukon, and Columbia, as large rivers draining wet portions of the continent, dominate total river flow to the surrounding oceans. The Mississippi, Mackenzie, and Yukon also transport much of the sediment moved to the oceans by rivers, along with the

during the autumn and winter. Rivers crossing the folded sedimentary rocks of the Appalachians have long stretches formed in bedrock, and more resistant rock types produce steeper channels with shorter meander lengths. Channels in the glaciated uplands toward the north are likely to have steep profiles and bouldery beds where they flow across glacial sediments. Once a river reaches the coastal plains, the channel is likely to meander broadly, forming complex wetland topography with natural levees, back swamps, and oxbow lakes. Rivers north of the St. Lawrence drain an area of lower topographic relief with numerous lakes. These rivers reach peak flow during the summer.

Rivers of this region mostly drain densely vegetated catchment areas where forest cover, well-developed soils, and lower intensity rainfalls promote subsurface flow down hillslopes and keep suspended sediment

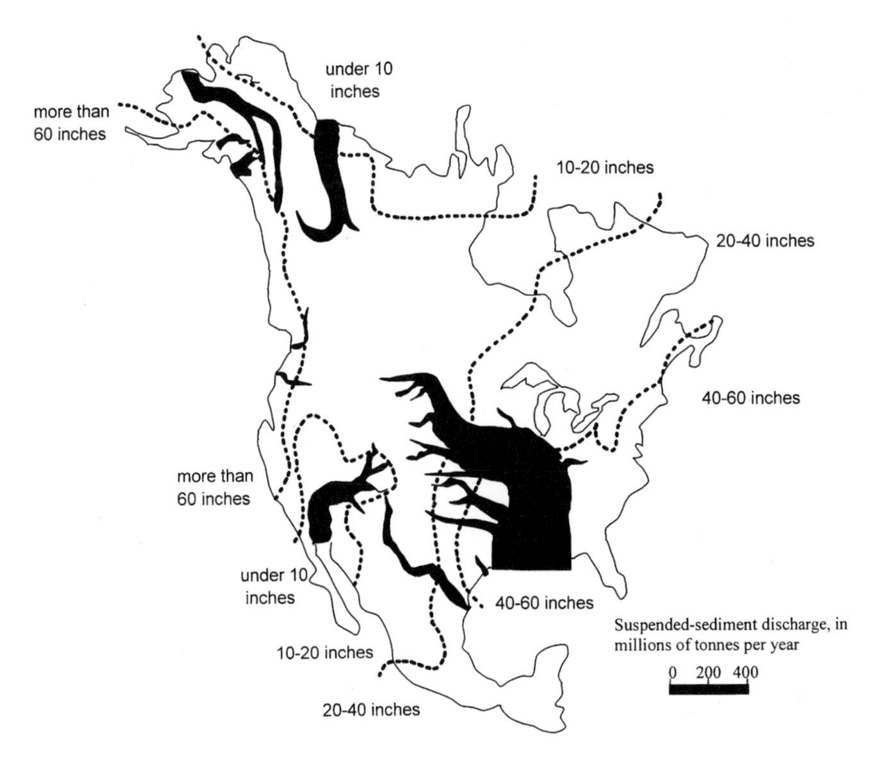

Colorado and Rio Grande. Note that sediment in the Mississippi River basin comes mainly from the western tributary rivers, such as the Missouri River, which drain the Rocky Mountains. (After R. H. Meade et al., 1990, Movement and storage of sediment in rivers of the United States and Canada, in M. G. Wolman and H. C. Riggs, eds., *Surface water hydrology*, Geological Society of America, Boulder, Colo., pp. 255–80.)

levels in the rivers relatively low. Exceptions occur when a dissipating hurricane crosses inland over the East Coast of the United States, bringing torrential rainfalls that trigger debris flows and landslides on hillslopes and flooding on rivers. Otherwise, rainfall is spread fairly evenly throughout the year. Flow variability on the rivers is relatively low, and floods may occur at any time of the year.

Rivers of the southern portion of this region share with rivers of the Lower Mississippi region the distinction of having both the greatest species richness (more than three hundred species) and the highest number of endemic species (approximately one hundred species) of aquatic organisms in North America. Endemic species are native to a particular region and are found only there. Species of fish, crayfish, and mussels are all abundant in eastern and southern rivers. Rivers of these

The Cheat River in West Virginia, of the Northeast and East-Central region, has cut a deep canyon along portions of its length. Along this river, which was south of the area of Pleistocene-age continental glaciation, the large boulders in the river come from rockfall along the canyon walls. Farther north, many rivers with catchment areas covered by glacial ice during the Pleistocene have a supply of large, rounded boulders that were left behind as the ice sheet melted. These boulders may be too large to be moved by the contemporary river.

regions host highly diverse organisms because glaciation did not disrupt or obliterate southern river systems, as it did in the north. Aquatic organisms in southern rivers thus had longer time spans over which to evolve into distinct species in the diverse habitats provided by a variety of geology, topography, and climate. Although contemporary fish stocks are much lower than those present at the time of European contact, the rivers of eastern North America still host a wide variety of anadromous fish that spend some part of their life cycle in the Atlantic Ocean. These fish include American shad (*Alosa sapidissima*), alewife (*Alosa pseudo-*

Durham Creek, North Carolina, is a blackwater stream in the Lower Mississippi region with clear water stained by organic acids from decaying vegetation. The river meanders across the coastal plain. (Courtesy of Brian Bledsoe.)

harengus), Atlantic salmon (*Salmo salar*), striped bass (*Morone saxatili*), and sturgeon (*Acipenser spp.*).

Rivers of the Lower Mississippi Region

Rivers of the Lower Mississippi region flow south to the Gulf of Mexico from the southwestern portion of the Appalachian Mountains and the eastern edges of the great interior plains. As they approach the Gulf, the rivers meander broadly across the flat plain created by thousands of years of river deposition. Peak flow occurs in autumn. The Mississippi River, also known as the Big Muddy, carries a large load of suspended sediment from the drylands in the western portion of its drainage.

Although now extensively channelized and confined by levees, rivers of this region historically created extensive bottomlands that flooded for several months each year. Widespread flooding was enhanced by features such as a seventy-five-mile-long logjam present on the Red River of Louisiana until European Americans blasted and destroyed it in 1834–35 and again in 1874. At smaller scales, broad, flat valley bottoms and abundant rains created the famous swamps and bayous where cypress and tupelo trees developed buttress roots in response to months of inundation and an otter leaves a trail of bubbles on the water surface as it carries crayfish back to its pups.

Rivers crossing the coastal plain of both this region and the Northeast and East-Central region may be blackwater rivers. These distinctive-looking rivers with sandy beds have clear but dark-colored water that is stained by dissolved organic matter including tannic acid from decaying vegetation. The water is acidic and low in dissolved nutrients and often enters the river after moving slowly through swamps and wetlands.

The Lower Mississippi region has served as an ecological refuge during times of glaciation and a source for organisms migrating to newly deglaciated regions. Unique to the Lower Mississippi are two extant species of paddlefish, ancestral catostomid suckers, the genus *Alligator*, and giant aquatic salamanders.

Rivers of the Central Region

Rivers in the northern part of the Central region drain northeast to Hudson Bay; those to the south drain east to the Mississippi River or directly to the Gulf of Mexico. Within the United States, these rivers of the dry continental interior flow year-round if they carry snowmelt from the Rocky Mountains or are supplied by groundwater. Rivers of the tallgrass prairies along the eastern half of the Central region often start from prairie potholes and springs. The rivers are likely to be ephemeral, flowing only a short time after snowmelt or rainfall if they head on the dry interior plains in the western half of the region and are supplied mainly by surface runoff. Peak flow occurs during spring or summer when melting snows and intense rains carry large volumes of fine sediment into the rivers. A prairie stream can resemble café au lait after a summer thunderstorm.

Early European explorers and settlers described these turbid plains rivers as being a mile wide and an inch thick. The broad, shallow rivers braiding or meandering eastward stitch together the Rockies, the shortgrass prairies at their feet, and the tallgrass prairies farther east. The river corridors formed lines of east-west travel for organisms from birds and fish to mammals throughout geologic and historic times. The eyes of humans traveling across the flat, grassy plains were drawn to the groves of cottonwoods lining the rivers.

The Platte River of the Central region in Nebraska is a broad, shallow stream with vegetated islands. During low flow the river splits into multiple channels; during high flow it floods the adjacent lowlands. Nineteenth-century visitors to the Platte described it as being a mile wide and an inch deep.

Rivers of the Central region have a rich fish fauna with many species. Dominant families are minnows, darters, topminnows and killfishes, catfishes, suckers, and sunfishes and black basses. Species adapted to large, main-channel environments in this region include the pallid (*Scaphirhynchus albus*) and the shovelnose (*Scaphirhynchus platorhynchus*) sturgeon. Few species are endemic to the Central region, but the most characteristic fish fauna are suites of species adapted to shallow, warm river and creek environments.

Rivers of the Southwestern Canyon Region

As the Colorado Plateau of Arizona, New Mexico, Colorado, and Utah is gradually lifted up by forces in the Earth's interior, the channels of the Colorado River drainage system cut downward to keep pace with the uplift. This river incision creates deep, sheer-walled canyons as abrupt as knife gashes in the high, dry plateaus of the southwestern United States. The large rivers, supplied by snowmelt, flow year-round and peak in

This river of the Southwestern Canyon region has cut into volcanic rocks to form the deep Boulder Canyon in central Arizona. The river has very low flow except after summer thunderstorms.

early summer. The smaller rivers are mostly ephemeral, but even some of these have incised deep canyons. Most of the rivers are extremely turbid, for they drain erodible sedimentary rocks in a dry climate where sudden rains can flush sediment down small rivers in debris flows and flash floods.

These rivers flowing through hot desert create oases for plants and animals. Many of the native fish are endemic species that are restricted to the Colorado River basin, having evolved to cope with the relatively warm, turbid water and the seasonal fluctuations. Much of the time, the clear weather of this region and the lack of vegetation promote an appearance of timelessness, as though the rivers and their landscape have always appeared as they do at this moment. But the cumulative effect of many flash floods and debris flows spread over decades actually creates fairly rapid change in these canyons, particularly where the flow has been altered by the construction of dams or the diversion of water from river channels.

The Colorado River and its large tributaries historically hosted an extraordinary assemblage of large-river fish species. The humpback chub (*Gila cypha*), bonytail (*Gila elegans*), Colorado pikeminnow (*Ptychocheilus lucius*), roundtail chub (*Gila robusta*), and razorback sucker (*Xyrauchen texanus*) all have morphological adaptations for life in turbid, fast-flowing waters. Most of these species are now threatened or endangered as a result of flow regulation from dams, water withdrawals, and introduced nonnative species.

Rivers of the Western Cordilleran Region

The Western Cordilleran region, stretching from southern California to Alaska, includes a tremendous amount of diversity, but the rivers share the characteristics of steep, mountain streams as they flow from the backbone of North America. The rivers of this region make a relatively short journey west to the Pacific Ocean or, if they originate on the eastern side of the Continental Divide, a longer journey east to the Atlantic Ocean. Many of the rivers start as staircase-like channels with alternating steps and small plunge pools. As channel slope decreases downstream, pools and riffles appear, and the river may become meandering once it crosses the coastal plain. The upper reaches of the river are likely punctuated by the drama of debris flows and landslides as the shallow soils of the steep headwater terrain are saturated by rainfall or snowmelt. Except for the southernmost part of this region, the rivers drain densely forested catchments, and woody debris in the channel provides important habitat diversity and stability.

The timing of peak discharge varies with latitude. Rivers have winter rainfall peaks in the southern portion of the region, winter rainfall and spring snowmelt peaks farther north, and summer snowmelt peaks in northern Canada and Alaska. Suspended sediment load is moderate, but the mountainous portions of the rivers receive abundant coarse sediment from adjacent hillslopes. Where glaciers are present, the large volumes of water and sediment released by the glaciers create braided streams. In the southern portion of this region, extremely dry condi-

The Bear River of northwestern California is in the Western Cordilleran region and has pools and riffles formed in cobbles and boulders. It flows at a moderately steep gradient through densely forested mountains.

tions can produce internal drainages that end in a saline lake or a desert basin without reaching an ocean.

Rivers reaching the Pacific Ocean from this region historically had large runs of salmon and trout. Principal species included pink (*Oncorhynchus gorbuscha*), chum (*O. keta*), sockeye (*O. nerka*), chinook (*O. tshawytscha*), and coho (*O. kisutch*) salmon, as well as cutthroat (*O. clarki*) and steelhead (*O. mykiss*) trout. Many of the anadromous fish runs no longer exist because of the combined effects of overfishing, introduced fish species, flow regulation and dams built along the rivers, and habitat destruction of spawning gravels and pools in the rivers caused by increased sediment from logging, mining, and other land uses.

Rivers of the Arctic Region

Rivers of the Arctic region drain north to the Arctic Ocean, carrying large amounts of fine sediment released into the flow from melting glaciers and as partially frozen streambanks collapse during the brief summer. Flow peaks during summer as the snowpack melts. Ice jams may

This river of the Arctic region drains a glacier in Denali National Park in central Alaska. The river is braided and shifts back and forth repeatedly across its broad floodplain as pulses of water and sediment are released by the upstream glacier.

dramatically elevate local water level and cause pulsed flooding during the spring ice breakout, but this effect is becoming less pronounced as global warming changes the historical breakout to a swifter, less dramatic meltout.

This region includes the immense Yukon and Mackenzie Rivers, which form channels that stretch to the horizon in the flat vastness of the Arctic plains. The region also includes smaller channels that may be ephemeral, or formed on the snow, giving way to poorly drained wetlands later in the season. Rivers are likely to be meandering or to have multiple channels that split and rejoin in an anabranching pattern.

Like those in the Western Cordilleran region, rivers of the Arctic region host anadromous salmon that migrate long distances between the ocean and their spawning grounds. The region has a low number of endemic species.

Beyond these factual descriptions of the rivers of the United States, there is also the poetry of their names, reflecting long histories of Native American and European American presence. In the Northeast and East-Central region flow the Androscoggin, the Kennebec, the Penobscot, the James, the Rappahannock, the Roanoke, the Altamaha, the Ogeechee, and the Suwanee. Into the Lower Mississippi region flow the Cumberland, the Tombigbee, the Yazoo, and the Atchafalaya. East from the Rockies flow the Belle Fourche, the Cheyenne, the Loup, the Platte, the Smoky Hill, the Purgatoire, the Apishapa, the Cimarron, and the Concho. Creating the southwestern canyons are the Fremont, the San Rafael, the San Pedro, the Gila, the Verde, the Hassayampa, the Moenkopi, and the mighty Colorado. From the Western Cordilleran region flow the San Joaquin, the Tuolumne, the Sweetwater, the Willamette, the Skagit, the Puyallup, the Nisqually, the Quinault, the Klickitat, the Skeena, and the Nushagak. And in the far north lie the Yukon, the Koyukuk, the Tanana, and the Anaktuvak. These names are also our heritage.

The River Ecosystem

An ecosystem consists of all the organisms in a community and the associated nonliving environmental factors with which they interact. A river ecosystem includes an upland zone, riverside zone, aquatic zone, and hyporheic zone. The upland zone lies beyond the elevation of flooding but serves as a source for water and sediment, as well as organic materials, entering the river. Animals such as deer or grouse that spend much of their life in the upland zone may depend on the river for food or water seasonally or during certain portions of their life cycles. These visitors from the upland zone can play an important part in structuring aquatic communities or in moving nutrients between the river and the uplands, as when grizzlies and eagles feast on spawning salmon.

The riverside zone beside the river is inhabited by plants and animals adapted to periodic flooding, fluctuating water tables, and disturbances as the river channel moves laterally. The riverside zone becomes a part of the aquatic habitat during periods of overbank flooding, and contributes nutrients to the river throughout the year. Many plants and

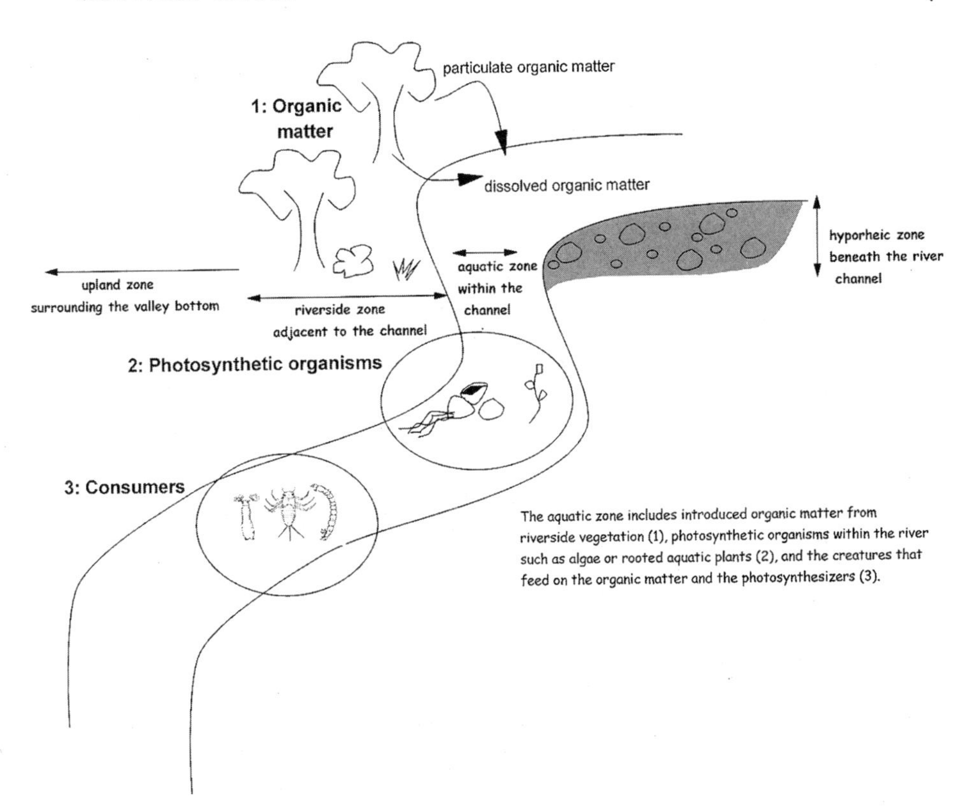

Schematic diagram of a river ecosystem.

animals have structured their life cycles and sites of residence around regular seasonal flooding and drying of the riverside zone. Plant communities along the coastal-plain streams of Virginia, for example, reflect the typical duration of the annual inundation for each surface: the wettest areas are dominated by shrub-scrub wetlands, whereas progressively drier areas support Cypress-Tupelo forests, and bottomland hardwoods with oak, hickory and ash species. In the Colorado River basin, endangered razorback suckers (*Xyrauchen texanus*) spawn each spring at sites where the river leaves a narrow canyon and spreads out across the bottomlands during peak flow. Timing and location of spawning indicate that annual flood peaks play an important role in enhancing survival of newly-hatched suckers by providing access to warm, food-rich floodplain habitat.

The aquatic zone within a river channel includes introduced organic matter and photosynthetic organisms that serve as the base of the river food web. Organic matter ranges from large logs down to fine particles less than a fraction of an inch, and may even enter the river in dissolved form. Photosynthetic organisms cover an equally broad range, from microscopic algae, bacteria and protists suspended in the water of large rivers to readily visible filamentous algae, mosses, and vascular plants that fill the role of photosynthesizers in mid-sized rivers.

Also inhabiting the aquatic zone are the consumers that feed on organic matter and photosynthesizers, or on other consumers. This diverse group includes invertebrates—microbes, aquatic insects, crustaceans and mollusks—as well as fish, amphibians, reptiles, birds, and mammals. Most aquatic organisms have adapted to very specialized conditions of streambed material and water velocity, temperature, and chemistry. Stone flies, for example, prefer the swift, cold, oxygenated waters of a rocky riffle in a mountain stream, whereas blackflies and net-spinning caddis flies can be found in the warm, slow flow and muddy bottom of a pool in a lowland stream. Stream ecologists have subdivided aquatic insects and fish into categories that reflect feeding preferences. Insects include shredders that feed on nonwoody coarse organic matter; shredder/gougers that feed on woody coarse organic matter; filterer-collectors that feed on fine organic matter and microscopic creatures moving in suspension through the water column; collector-gatherers that feed on fine organic matter and microscopic creatures deposited on the riverbed; grazers that feed on algae growing on the streambed cobbles; and predators that feed on rooted aquatic plants and animal prey. Similarly, fish include piscivores that feed on other fish; benthic invertebrate feeders eating bottom-dwelling insects; surface and water column feeders eating higher in the water column; generalized invertebrate feeders; planktivores that eat plankton; herbivore-detrivores that feed on plants and detritus; omnivores that eat a variety of things; and parasites.

Below the river and valley bottom lies the hyporheic zone, a portion of the groundwater system that is directly connected to the aquatic zone at sites of upwelling into the river and downwelling into the subsurface. Many aquatic insects live a portion of their life cycle in the hyporheic zone, and nutrient exchange between river and hyporheic waters influences aquatic communities.

Organisms living within the river ecosystems of the United States have evolved a wonderful diversity of life strategies to match the diversity of available habitats. Within the same river region, in this case the Northeast and East-Central region, anadromous fish such as shad migrate hundreds of miles to scatter thousands of eggs and sperm during spawning, while darters, a type of perch, ascend a mile up tiny rivulets seeping from ephemeral ponds in meadows to attach a few eggs to submerged blades of grass. Mussels rely on other organisms, primarily fish, for transport. Female mussels brood thousands of larvae in specialized portions of their gills. Mature larvae are released into the water to find hosts, where they attach themselves to the gills, fins, or lips and remain encapsulated from days to months. Some mussel species are generalists able to use almost any fish that happens by; others are specialists that need a certain type of fish, such as a darter. Eventually the larvae break out of their capsules and settle to the stream bottom, where they spend the rest of their lives embedded in the sediment, filtering food from the water with their gills. The variety of adaptive strategies present among the organisms living in each river reflects the richness of river evolution.

Humans and Rivers

River channel self-adjustment and variability may create hazards for humans in the form of floods, bank erosion, or bed scour around bridge piers. The most common goal of river engineering is to stabilize a river channel—to make it "behave," so that the water is always contained within the channel and the channel itself does not move or alter substantially. Humans try to treat rivers as canals, and diversity and variability are unwelcome challenges. But the variability of a natural river creates a diversity of habitats that support an assortment of aquatic organisms in the channel and riverside organisms in the floodplain. Many of these organisms are adapted to or require the disturbances associated with a natural river. As mentioned earlier, the seedlings of cottonwoods may germinate only after a flood has removed some of the older trees and freshly deposited a sandbar. The more diverse the physical habitats along a river, the greater the diversity of organisms occupying those habitats. A channel with pools and riffles may have grazers feeding on the algae growing on the riffle cobbles in the streambed. Along the stream margins, shredders feed on accumulated leaf litter. In the pools, collector-gatherers feed on fine organic material.

A diverse ecosystem is more healthy and stable because it is able to absorb disturbances such as floods, or even toxic chemical spills, without completely breaking down and suffering massive species loss. Some species may be temporarily removed from a site by the disturbance, but others will survive. Over time, the species sensitive to disturbance will recolonize a site as water temperature, streambed composition, or other conditions return to predisturbance levels. Because a drainage canal or a channelized river has simpler river form, flow hydraulics, and sediment transport than a natural river, the canal or altered river supports a less diverse community of organisms.

Because the organisms inhabiting a river ecosystem adapted through evolution to the specific physical and chemical conditions of that river, any human alteration of the river affects the organisms. Temporary, local, or modest impacts lead to a shift in the spatial distribution of organisms, or to a change in the species composition of the river community. For example, construction of a low dam may cause sediment accumulation and loss of naturally occurring pools for a mile upstream, with associated changes in fish and insect communities. Prolonged, extensive, or severe impacts lead to loss of species and decreased biological diversity within the river. Leaching of heavy metals from mine tailings may contaminate water and streambed sediments for fifty miles downstream and destroy fish communities for decades.

Humans affect river ecosystems directly through activities that occur within the river channel. Channelization, sand and gravel mining, or fishing each alter the form or function of a river. Indirect impacts occur elsewhere in the drainage basin but affect the movement of water, sediment, or other materials such as toxic chemicals into the river. Timber harvest in the uplands or urbanization within the drainage basin each alter the amount and timing of water, sediment, nutrients, and chemicals moving from the uplands into the river.

The history of settlement patterns and socioeconomic conditions superimposed various types of human land-use impacts on the six river regions of the United States. These impacts varied through time, but river ecosystems preserve a memory of them. No river in the United States has escaped the imprint of humans, as becomes apparent if you know what to look for.

Examples of Direct and Indirect Human Impacts to River Ecosystems

Direct

Flow regulation

Dams and diversions alter the magnitude and timing of water movement along a river, thus affecting sediment transport, river stability and form, water temperature, and the availability of dissolved oxygen, nutrients, and habitat. Dams also impede upstream-downstream movement of in-channel organisms such as fish and the seeds of riverside vegetation. If flood peaks are reduced by dams or diversions, the lateral connectivity between the channel and its floodplain is reduced.

Channelization and levees

Channelization reduces river habitat diversity and, along with levees, alters the magnitude and timing of water movement along a channel by preventing overbank flow of water. By reducing or eliminating the exchange of water and sediment between the river and the floodplain, channelization and levees also change the availability of nutrients and habitat.

In-channel mining

Mining of placer metals or sand and gravel may alter river form, boundary resistance, and sediment transport. The usual results are an increase in fine sediment transport and water turbidity, an increase in channel mobility, and a loss of pool volume and habitat diversity.

Beaver trapping

Beavers build low dams of sediment and wood across river channels. These dams slow the passage of flood flows, store sediment, locally elevate the water surface and water table, and increase habitat diversity. When the beaver are removed and the dams fall into disrepair, channel erosion, increased sediment transport, flashier flood flows, and loss of habitat diversity may result.

channel cross section

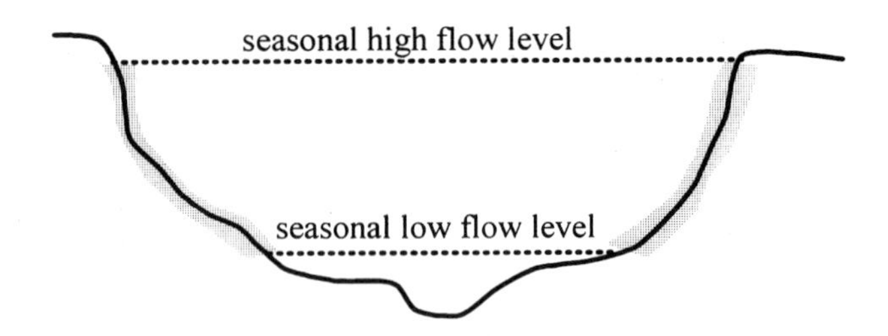

Natural River
snowmelt hydrograph with gradual rise and fall of flow

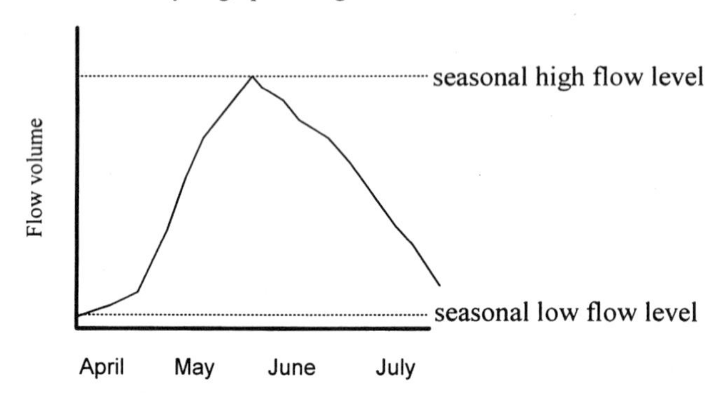

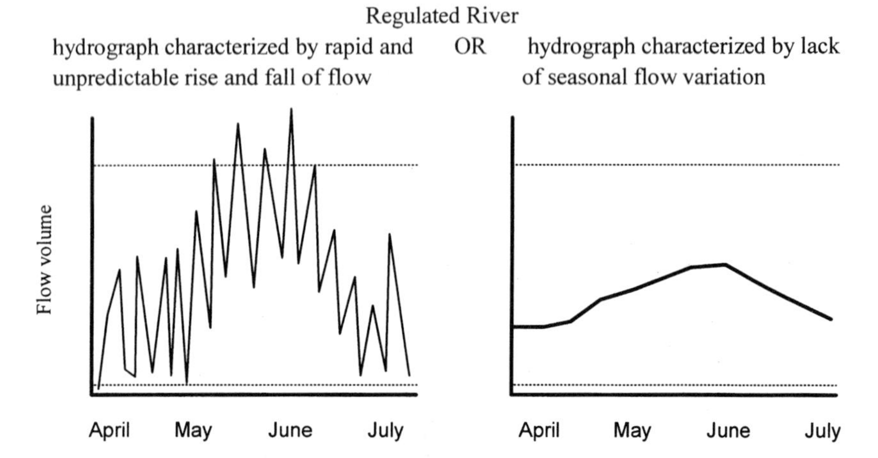

Impacts associated with flow regulation. Rapid and unpredictable rise and fall of water in the zone along the river margins can expose aquatic organisms to air, causing them to dry or freeze, or can flood organisms, reducing sunlight penetration, photosynthesis, and water temperature. Reduced seasonal variability in flow either can eliminate thermal or chemical cues on which organisms base their life cycles or can eliminate access to seasonal habitat.

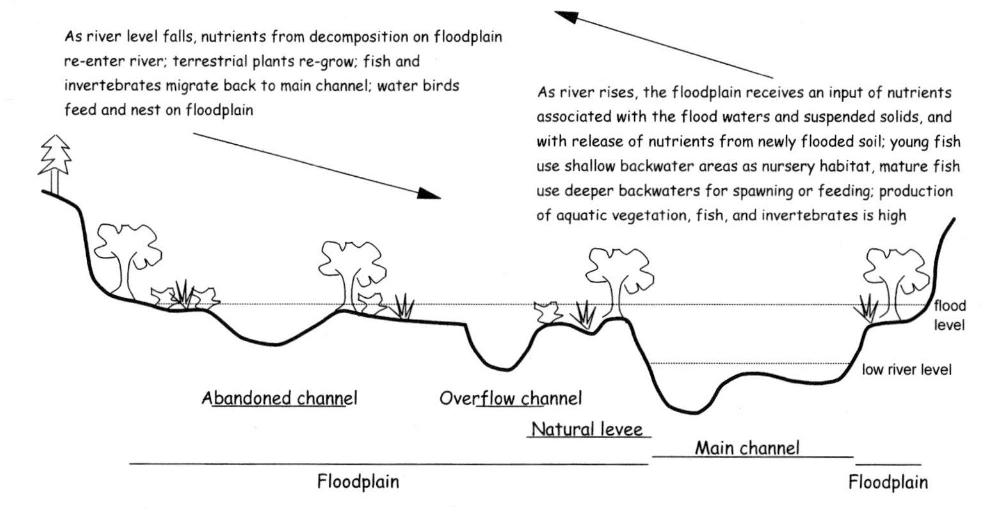

Channelization and levees. River-floodplain connections in an unimpaired river create the seasonal fluctuations shown here; reduction of overbank flooding reduces these connections.

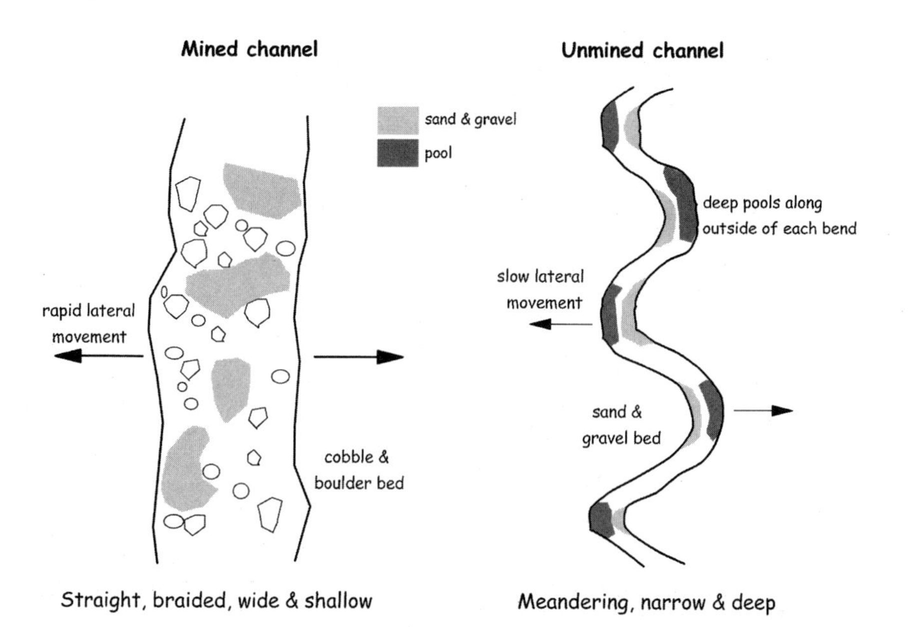

Schematic illustration of changes in river characteristics a century after placer mining on the headwaters of the South Platte River in Colorado.

Beaver dams along a river in West Virginia create ponds that trap sediment and organic material, creating more diverse in-channel and riverside habitat.

Wastewater effluent

Treated wastewater effluent may account for as much as 100 percent of stream flow downstream from major urban areas. This effluent may be the primary source of nitrate, ammonia, and phosphorus to streams and contributes heavy metals, PCBs, and other toxic chemical compounds.

Floodplain encroachment

Filling seasonal wetlands and constructing building and transportation corridors on the floodplain limits lateral movement by the river, destroys floodplain habitat, and reduces channel-floodplain exchanges of water, sediment, nutrients, and organisms.

Snagging and removal of wood

Wood in channels creates habitat diversity by directing flow that forms scour pools in the streambed and banks, by ponding water and storing fine sediment and organic detritus, and by providing overhead cover and diverse stream substrate for in-channel organisms. Some large rivers in the United States historically had huge, persistent logjams that ponded water for tens of miles upstream, creating extensive floodplain wetlands. Removal of wood to facilitate downstream movement of water and sediment reduces channel stability and diversity.

Indirect

Timber harvest

Removal of slope vegetation and construction of roads in association with timber harvest usually increase water yield from hillslopes to rivers and, by destabilizing the hillslopes, dramatically increase sediment yield to rivers. Increases in sediment yield in turn increase water turbidity, fill pools, destabilize channels, and decrease habitat diversity.

Agriculture

Planting crops and grazing domestic livestock usually result in the reduction of natural vegetation cover and an increase in sediment yield from hillslopes. Agriculture may also create nonpoint sources of nitrogen from animal waste or pesticides and herbicides that eventually enter the river from surface runoff or groundwater inflow. Chemicals used as pesticides may be particularly long-lasting; a 1992–95 sampling program along the South Platte River basin of Colorado found residues of DDT in riverbed sediments and fish tissues although this chemical was banned from use in 1972.

Urbanization

The increase in pavement associated with urbanization dramatically decreases sediment yield and increases water yield to a river. This usually results in channel erosion and associated channel instability and downstream increases in sediment transport. Runoff entering a river from urban areas also has contaminants ranging from household pesticides and gasoline to the effluent of wastewater treatment plants. These contaminants can severely stress aquatic and riverside organisms and alter the species composition of river ecosystems.

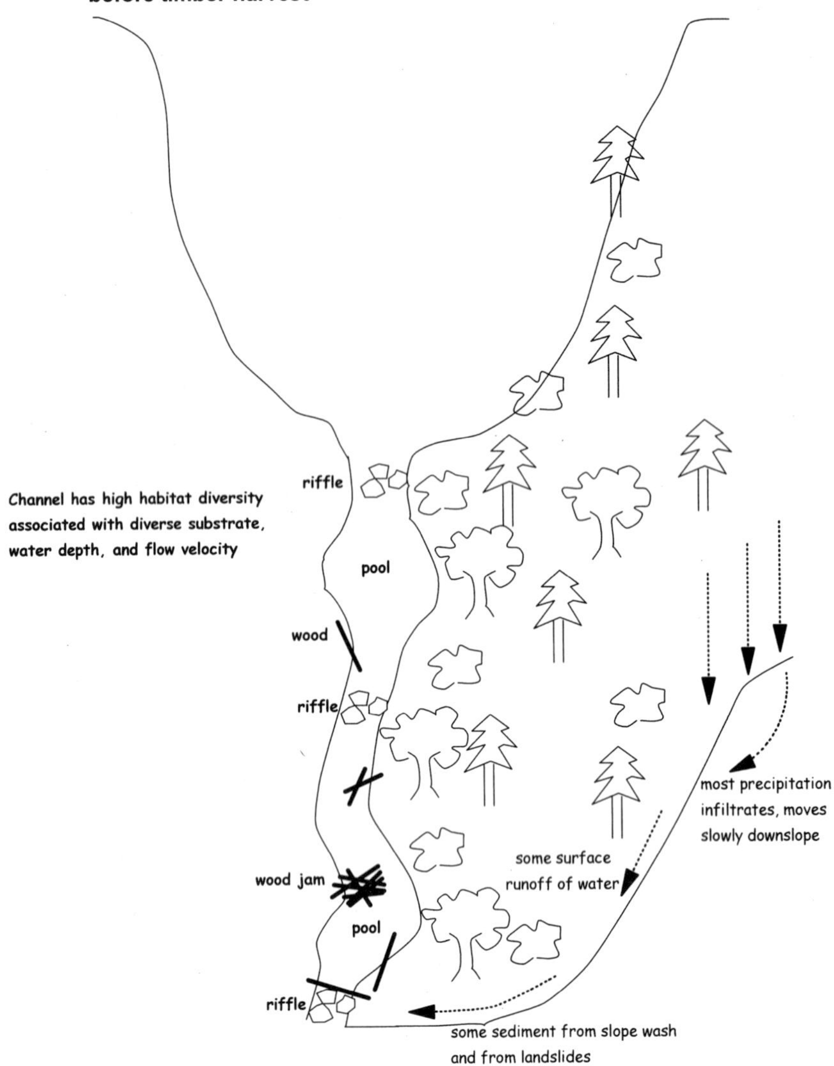

Schematic upstream view of a stream and adjacent hillslopes before and after timber harvest.

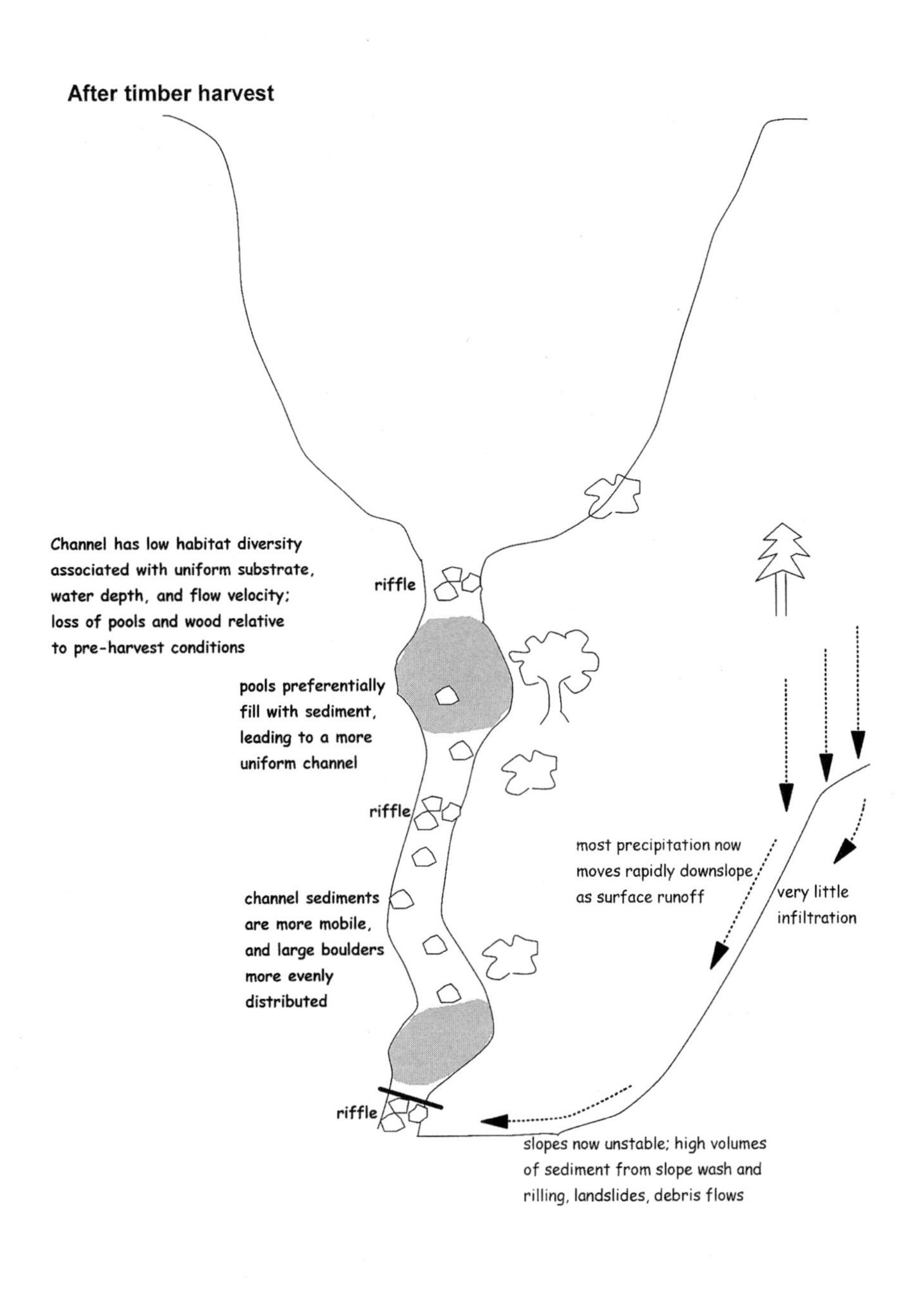
After timber harvest
Channel has low habitat diversity associated with uniform substrate, water depth, and flow velocity; loss of pools and wood relative to pre-harvest conditions
riffle
pools preferentially fill with sediment, leading to a more uniform channel
riffle
channel sediments are more mobile, and large boulders more evenly distributed
riffle
most precipitation now moves rapidly downslope as surface runoff
very little infiltration
slopes now unstable; high volumes of sediment from slope wash and rilling, landslides, debris flows

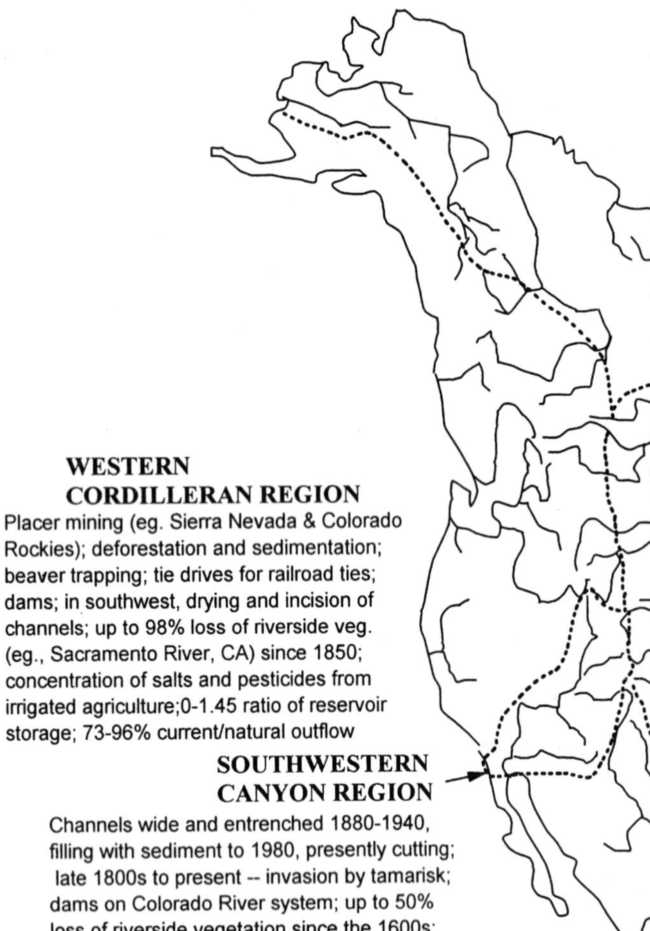

Examples of human impacts to rivers by region within North America.

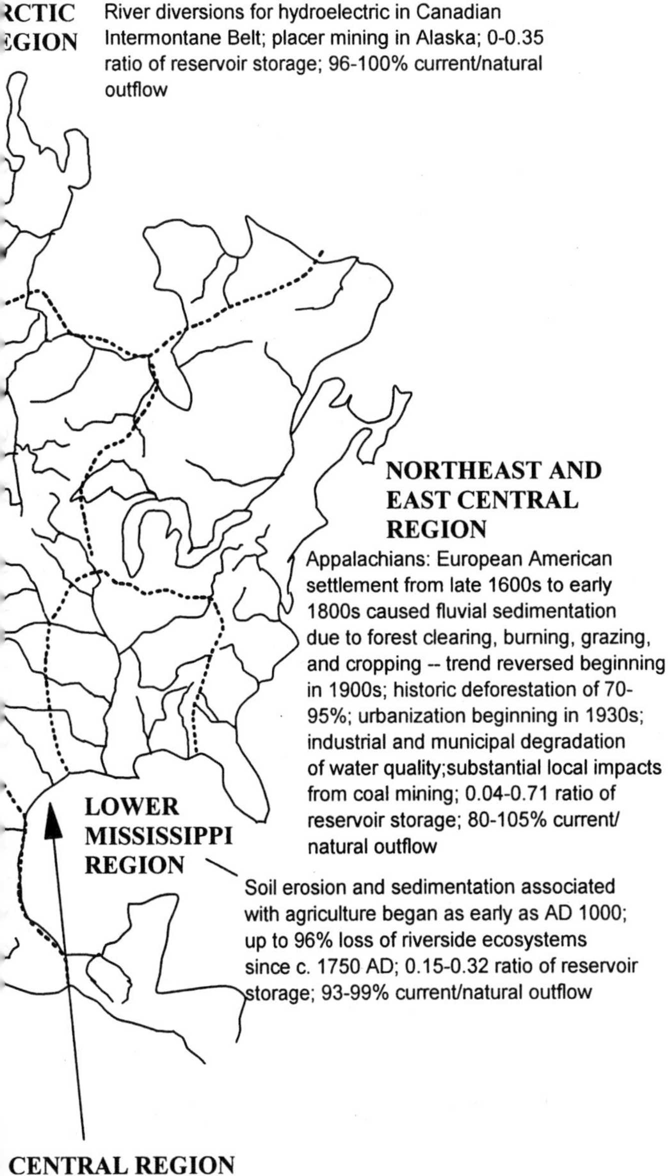

CENTRAL REGION

Major channel change since late 1800s resulting from irrigation diversions, reservoirs, and flow regulation; more than 50% loss of riverside ecosystems along some channels since 1880; degradation of water quality from agricultural runoff (pesticides, animal wastes);0.14-0.19 ratio of reservoir storage; 93-99% current/natural outflow

Chapter 3

Conquering a New World

Pioneer Impacts

Humans reached North America at least twelve thousand years ago. The first people likely migrated south after crossing the Bering Strait region from northern Asia. They may have come as early as forty thousand years ago and moved southward along the coastal region or through the interior. What we are certain of is that by twelve thousand years ago people were living throughout the length and breadth of the Americas.[1]

These earliest human inhabitants were hunter-gatherers. They modified the physical and biological environments locally through practices such as setting wildfires or building fish weirs in rivers. Native American burning and alteration of the landscape was pervasive and important in shaping plant communities, but there is no evidence that these peoples substantially modified the forms or processes of rivers.[2]

Sedentary agricultural communities first appear in the archeological record of North America more than two thousand years ago. As with agricultural peoples elsewhere in the world, these first farmers affected rivers more than their hunter-gatherer predecessors. The Hohokam of central Arizona, for example, developed an extensive system of more than twelve hundred miles of irrigation canals diverting water from the Salt River between approximately A.D. 200 and 1450. Native Americans in the eastern woodlands farmed along river corridors. Their domestication of seed plants began in approximately 2000 B.C. Between 250 B.C. and A.D. 100 they developed more extensive food production systems, and between A.D. 800 and 1100 maize became their dominant crop. Much of this was swidden agriculture, where forested land was cleared with fire, planted for a few years, and then abandoned as the farmers burned a new plot. The great mound-building civilizations that sprang up by 1500 B.C. in the lower Mississippi River valley and spread north to the Ohio River valley by 500 B.C. were agricultural communities based on maize, beans, and other crops. As many as five thousand people lived in each of these communities. Seasonal use of oxbow lakes and other floodplain environments in the Mississippi River valley is indicated by ragweed (reflecting vegetation disturbance) and maize pollen in sediment cores. By A.D. 800–1100 maize was being grown from the Gulf of Mexico north to the Great Lakes and from the Atlantic coast west to the Great Plains. Half the animal protein consumed by people living along the Mississippi River valley came from backwater fish and waterfowl.[3]

Estimates of the pre-Columbian population of North America range widely, between two million and eighteen million people. In the drier regions farmers diverted water from rivers or modified dry channels with small dams to store water after a flash flood. In the wetter regions they burned native vegetation to create clearings for their crops, probably increasing sediment yields to local rivers in the process. Wildfire was extensively used, from the Seminoles hunting alligators with fire in Florida, to the Iroquois practicing swidden agriculture in New York, and to the Sioux and Apaches burning to improve grazing in the western United States. Stephen Pyne estimates that across North America, Native Americans had repeated, controlled surface burns on a cycle of one to three years, broken by occasional firestorms from escaped fires and during times of drought. These repeated burnings replaced dense forests with grassland, savannah, or open woodland and were critical in creating the tallgrass prairies of the eastern Great Plains. The fires also increased sediment and water yields to rivers. We know relatively little of how rivers responded to these changes, but population densities and thus land-use impacts were probably sufficiently low that most rivers were relatively stable at the time of European contact.[4]

Except for brief visits by the Vikings, initial European contacts along the eastern and southern borders of the present United States spread across one hundred years. The Spaniards began to explore the fringes of the eastern coast of North America in the 1520s. Ponce de Léon visited briefly in 1521 with eighty men, and the 1528 Narvaéz expedition left Cabeza de Vaca to travel for eight years from Florida west to Arizona and then south to Mexico City. Hernando de Soto wandered from Florida up into the southern Appalachians and then west to the Arkansas and Red

Rivers between 1539 and 1543, and Francisco Vásquez de Coronado traveled up from Mexico into the Rio Grande valley in 1541–42. The French may have reached the northeastern coast of North America before 1492, and they were certainly present by the late 1500s. Samuel de Champlain mapped and explored the Atlantic coast between 1604 and 1607, and English colonists arrived in 1607.[5]

Over the next three centuries the European Americans steadily spread outward from their initial points of entry, exploring ways to utilize the continent's natural resources. At the time European Americans reached this continent, their material technology was much more resource-intensive than that of the Native Americans. The European Americans were also a peripheral part of a global trading network that drew heavily on any new resources that they found to exploit. And the European Americans were backed by a huge reservoir of humanity ready to expand into new territories. All of these differences meant that European Americans almost immediately began to engage in activities that affected rivers and other natural resources much more directly than most of the activities of the Native Americans.

One of the first effects on rivers of the St. Lawrence drainage and New England was the trapping of beavers during the seventeenth century. Removal of the beavers caused their dams to fall into disrepair, allowing floods to pass more quickly down rivers, flushing the sediment stored in beaver ponds and decreasing the diversity and availability of habitat for plants and animals.[6]

The European Americans formed sedentary agricultural communities and immediately began to clear forest cover, plant crops, pasture grazing animals, and build roads. Removal of forest cover accelerates the movement of water, sediment, and nutrients downslope into river channels. The aboveground portion of vegetation reduces the impact of raindrops. Vegetation also shelters and shades snow so that it melts more slowly. And plants return some moisture directly back to the atmosphere through evapotranspiration of water from plant surfaces. The leaves and twigs falling from plants create a layer of decaying organic material that increases the ability of the soil to absorb water. The roots of plants both bind the soil in place to make a hillslope more stable and increase the ability of the soil to slowly absorb and release water.

When vegetation cover is removed, all of these stabilizing and filtering effects are lost. Sediment yield from deforested slopes increases first and remains higher for up to ten years until vegetation cover begins to regrow. Water yield from the slopes remains elevated for up to thirty years as the hillslope adjusts to persistent changes in water infiltration and the absence of a network of plant roots. During the adjustment period after deforestation, sediment and water yields can double. Such increases in turn affect rivers by increasing the turbidity of the water, filling pools with silt and sand, increasing erosion of the streambed and banks, and decreasing habitat diversity and stability. Deforestation began during the period of pioneer settlement and continued through the development of commercial activities. Ninety-six percent of the original virgin forests of the northeastern and central states were gone by the 1920s. Throughout the conterminous United States, 98 percent of the virgin forest was gone by 1990. The effects of deforestation are thus ubiquitous.[7]

One of the first studies documenting the effects of forest clearing and subsequent agriculture focused on the Piedmont of Maryland. The 1967 study indicated that sediment yields from forested areas during the era before European American farming were less than one hundred tons per square mile per year. Yields from the same region increased to three hundred to eight hundred tons per square mile per year with the advent of agriculture. Subsequent studies demonstrated similar increases in sediment yield associated with farming throughout the United States.[8]

The receiving river might respond dramatically to the increased sediment yield. The Alcovy River, draining approximately five hundred square miles on the Georgia Piedmont, was surrounded by relatively dry, fertile bottomlands prized by European Americans settling the watershed starting in 1814. But sediment from erosion of newly cleared and poorly farmed uplands soon filled the streams. By the time of the Civil War, the river bottomlands were an extensive swamp that continues to accumulate sediment. Portions of the riverbed rose six and one-half feet between 1882 and 1992.[9]

As the agricultural front moved west across the United States, farmlands in the East grew back into woodlands, and sediment yields again declined. Several studies indicate that the effects of agriculture can be reduced when land is removed from production. The percentage of cultivated land in the seven and one-half square miles of the Goodwin Creek watershed in Mississippi decreased from 26 percent in 1982 to 12 percent in 1990. During this period the concentration of silt and clay in the creek decreased by 62 percent and the concentration of sand by 66 percent as the source of readily eroded sediment was removed.[10]

At the same time that European Americans began agriculture in a region, they built small dams along rivers to provide power for sawmills and gristmills. These dams altered water flow, sediment movement, and river form. As early as 1691, English colonists in Maine promised to give their Native American neighbors corn as compensation for ruining their fishing grounds. The local sawmill produced so much sawdust that, when the dust was shoveled into the water beyond the mill wheel, it drove away the fish. Throughout New England, New Jersey, New York, and Pennsylvania, sawmill waste locally fouled streams. More widespread and persistent were the soil erosion and stream siltation caused by deforestation. By 1750, New England streams that once swarmed with seasonal migrations of herring were thick with mud and empty of fish.[11]

Land-use changes including agriculture and forest clearing had the potential to exacerbate naturally occurring floods. When the increased sediment supply produced by land use accumulated in rivers, channel capacity was reduced, and floods became more likely to spill beyond the channel and across the valley bottom. Where river flooding threatened a community, levees were built to reduce overbank flow, thus altering the river channel. Landowners in the low-lying areas around New Orleans, Louisiana, built discontinuous levees along the lower Mississippi River beginning in 1718.[12]

Rivers were also altered to enhance the passage of barges or flatboats, or the passage of logs being floated downstream to a mill. Shallow areas such as riffles or sandbars were dredged, and natural obstacles formed by boulders or accumulations of logs were blasted away. Dams built to intercept logs also blocked the migration of fish. A 1798 dam built to catch logs on the Connecticut River was the first mainstem dam on a major river in North America, and it blocked the migrations of American shad. The first snagboat—designed to remove submerged or partially submerged logs from a river channel—was built in 1829 to remove logs from the Ohio and Mississippi Rivers. More than eight hundred thousand snags were removed from the lower Mississippi alone during the next fifty years. Over time, snagging extended to rivers throughout the Southeast, Midwest, and Pacific Northwest. Cottonwood and sycamore snags cleared from the Mississippi and Red Rivers averaged more than five feet in diameter. Snags pulled from rivers in western Washington were up to seventeen feet in diameter. Historical records suggest that enormous logjams several miles in length were once present along rivers as diverse as the Red River in Louisiana and the Willamette River in Oregon.[13]

Once the larger rivers were cleared of snags and jams, smaller tributary streams were catastrophically cleared through splash damming. Splash dams were temporary low dams built across a stream to pond water and trap timber cut upstream, which was then floated downstream to collection booms and markets. When a splash dam filled with logs, a charge of dynamite destroyed the dam and sent a surge of logs downstream on a dam-break flood. Splash damming was ubiquitous in forested regions as diverse as New England, the Pacific Northwest, the intermountain West, and the Midwest during the nineteenth century. The flood of water and logs resulting from breaching of a splash dam flushed sediment and wood from the streambed and banks, leaving many channels scoured to bedrock. Studies of otherwise analogous streams with and without a history of splash damming indicate that those without splash damming have tens to hundreds of times more naturally occurring wood, as well as more and deeper pools.[14]

Human and animal wastes from rapidly growing communities went directly into rivers, changing nutrient levels and water chemistry. The changes in sediment and nutrients entering a river, and the building of levees that inhibited overbank flooding and the movement of organisms and nutrients between the river channel and its floodplain, severely affected river fish communities. The most direct effects of sediment on fish habitat come from changes in streambed grain-size distributions and from loss of channel volume. Sediment entering rivers as a result of human activities such as agriculture, lumbering, or mining is predominantly composed of sand, silt, and clay. This fine sediment fills the spaces between cobbles and gravels on riffles. Newly hatched young of many fish species retreat to coarse riffle bottoms for overwinter cover, and the fine sediment filling reduces or eliminates the spaces essential to these tiny fish. The fine sediment also preferentially accumulates in pools, reducing pool water depth. This reduced pool depth decreases the physical carrying capacity for juvenile and adult fish during summer growth periods.[15]

The indirect effects of sediment on fish arise from changes in the community of bottom-dwelling insects on which fish feed. Sediment interferes with the respiration of insects living in spaces between cobbles. Large loads of fine sediment overwhelm filtering insects such as some caddis fly larvae that use fine-meshed catchnets to trap drifting food particles. And fine sediment simply smothers physical habitat on the streambed, including the tiny spaces occupied by burrowing insects.[16]

The ability of stream organisms to recover from excess sediment depends on physical characteristics such as streambed slope, flow, and velocity, which control the rate at which sediment is flushed. Recovery also depends on biological characteristics. Insects with flying adults can renew populations more quickly than nonflying organisms such as mussels or crustaceans. Undisturbed upstream reaches may also replenish downstream populations as organisms drift downstream with the current.[17]

The discharge of fermentable organic wastes—human and animal sewage, sawdust from sawmills, and so forth—creates a biochemical oxygen demand as bacteria and microbes break down the organic material. The biochemical oxygen demand is a measure of the amount of oxygen the fermentable organic matter uses in undergoing decomposition. The demand will depend on water temperature, re-aeration of the water, concentration of the organic matter, and time the organic matter has been decomposing. Warmer temperatures, for example, produce faster decomposition, which requires more oxygen and produces lower levels of dissolved oxygen in the stream water immediately downstream from the point where the organic matter enters the river. Fish and other forms of aquatic life need some minimum level of dissolved oxygen in order to carry out respiration, with the specific minimum depending on the species. Where the biochemical oxygen demand exceeds the capacity of the stream to purify itself via re-aeration and complete decomposition of introduced organic matter, dissolved oxygen falls below critical levels, killing stream fish and invertebrates.[18]

A study on Cullowhee Creek, an Appalachian mountain stream in North Carolina, illustrates the effects of excess sediment and nutrients on stream organisms. The upstream portion of the creek drains an undisturbed forest. This segment of the creek had sixty-four invertebrate species. The next segment downstream, with excess sediment from logging and residential construction, had fifty species of invertebrates. The downstream-most portion of the creek, with additional contamination by nutrients from horse pastures, had only thirty-six invertebrate species. Diversity, density, and biomass of river organisms were all decreased by the sediment alone, but in the organically polluted reach, density and biomass of some tolerant organisms were increased, shifting the community composition.[19]

The impacts of excess sediment and nutrients in rivers throughout the United States occurred simultaneously with the increase of fishing that often accompanied the establishment of European American communities with greater population densities than Native American communities. Settlers along Pennsylvania's Schuylkill River fenced the river in various ways during the early 1700s to intercept the spring migration of American shad. (In contrast, the mainstem Delaware River is free of dams in part because of a 1783 agreement between Pennsylvania and New Jersey not to block the shad runs.) Although pioneer-era fishing generally did not deplete fish stocks below sustainable levels, the fish communities already stressed by pollution from sediment and sewage were especially vulnerable to commercial fishing and to the introduction of nonnative species of aquatic life that occurred later during the period dominated by commercial impacts.[20]

In addition to the impacts of relatively small, pioneer human communities, individuals or small groups might exploit river resources heavily. Beaver trappers spread across the intermountain West during the first decades of the nineteenth century, well before European Americans developed communities in the region. By the time John C. Frémont traveled through Colorado in 1842, it was rare to find an active beaver colony.[21]

Placer miners quickly followed the beaver trappers into the intermountain West, rushing to California's Sierra Nevada in 1849, the Comstock region of Nevada in 1850, and the Colorado Rockies and the Fraser River placers of British Columbia in 1859. Placer mining refers to the removal of precious metals, primarily gold, dispersed in sediments deposited by rivers. Placer miners literally tore apart streams in their eagerness to get at the precious metals mixed in with the other stream sediments.

As more people moved to the western United States and developed agricultural communities, the impacts that had occurred to rivers in the eastern and midwestern United States were repeated. Diversion of flow from river channels for agricultural irrigation also exacerbated

A creek in central Alaska along which contemporary placer mining is occurring. The streambed has been thoroughly worked over, decreasing the stability and diversity of the streambed and banks. Pools and riffles have been obliterated, and any wood present has been flushed downstream. Riverside vegetation has been removed, and the connection between the creek and its floodplain has been severed.

human impacts. Diversion began with miners trying to create a sufficient water supply to separate metals from streambed sediments in sluice boxes. With the advent of agriculture, temporary diversion structures for mining gave way to much more substantial ditches and tunnels that often moved water from one drainage basin to another, sometimes even across the Continental Divide. Flow diversions changed the amount and timing of flow in the river channels, sometimes altering water temperature and chemistry, channel form and stability, and habitat availability for plants and animals.[22]

Individuals or small groups of people undertook many of the pioneer activities. However, this did not mean that the local impacts to rivers were inconsequential. The individuals sometimes used a particular resource so thoroughly that the river ecosystem took decades to recover. European American fur trappers removed nearly every beaver along all the mountain tributaries of the South Platte River in Colorado between 1820 and 1840. Or, so many individuals might simultaneously enter a region that the cumulative effect was devastating even if maintained for only a few years. Between three thousand and five thousand people rushed to Gold Run Creek in Colorado during the first season after the discovery of placer gold in January 1859. Or, many small communities might each create a localized impact by activities that persisted for decades. Regardless of the form of the impact, small numbers of people, or people inhabiting a region for only a short time, sometimes substantially impacted regional river systems.[23]

This first phase of European American exploration and settlement of the continent was the pioneer era. A pioneer prepares the way for others to follow. The people of this era were conscious of themselves as pioneers. They thought in terms of conquering the wilderness for civilization and following their manifest destiny across the continent. They perceived the landscape as being pristine because they did not acknowledge the role of Native Americans in altering that landscape. This pristine world was often regarded as an adversary, threatening the ability of humans to live in settled communities by producing floods, blizzards, plagues of insects, or dangerous wild animals. At the same time, this source of great danger could be made to produce resources useful to humans. While human population density remained low and natural resources remained largely unexploited on a continental scale, few people conceived that their actions adversely affected natural systems in any substantial way. The natural world seemed to offer limitless potential sources for personal and community property and wealth, and the European Americans came from religious and cultural traditions that emphasized human dominion over the Earth. These people crossed oceans to reach this continent, and they believed in actively exploiting their surroundings. Individual and societal initiative were highly regarded, and there were few societal controls on individual resource use in this land of perceived endless abundance. On the contrary, the federal government encouraged resource use through policies embodied in the Federal Land Survey (1785) or the Homestead Act (1862). The pioneers collectively regarded the natural world as a challenging obstacle that their ingenuity and persistence could overcome and put to good use.

The pioneer era lasted more than three hundred years. From the first incursions along the continental margins, European Americans gradually moved into the interior. It was not until 1893 that Frederick Jackson Turner made his famous pronouncement that the frontier was closed. In fact, however, the frontier had already been closed for decades in parts of the eastern United States, for the pioneer era was asynchronous, tracking the movement and settlement of European Americans through the United States.[24]

Simultaneous with the pioneer exploitation of natural resources by individuals and small groups was the pioneer study of those resources by scientists. Much of this scientific effort focused on inventory and classification. Carolus Linnaeus had developed his taxonomic system for living organisms in the mid-1700s, and European scientists had been busily classifying organisms ever since. Each new specimen brought to European scientists from around the world was described and placed in a species, genus, and other categories that implied something about its relations to other living organisms. The Americas provided a reservoir of unknown species, and the first task was to categorize them. This inventory was then used to test speculations such as those of the French Comte de Buffon, who hypothesized that North American mammals would be smaller, feebler versions of those in Europe, or Thomas Jefferson, who countered that North American mammals would in fact be vigorous giants.[25]

Simultaneous with the biological inventory was an inventory of physical resources for determining the capacity of the landscape to support human settlement. Meriwether Lewis and William Clark, Zebulon Pike, Stephen Long, and the leaders of other government-sponsored nineteenth-century expeditions into the continental interior were charged with this inventory. In addition to charting routes for incoming European American settlers, the explorers were to note qualities of climate, soil, and vegetation; supplies of water; and the presence of minerals as indicators of the potential for agriculture, lumbering, or mining. Some of the first tasks of newly formed governmental agencies such as the Geological Survey were mapping and measuring the land's contours and physical resources.[26]

These government inventories often occurred at the same time that pioneering individuals or communities directly tested the landscape's potential by conducting agriculture or mining. Placer mining was among the activities that most directly and dramatically affected American rivers during the pioneer era, and a more detailed examination of the extent and intensity of placer mining in the United States provides a case study of pioneer impacts to rivers. These impacts started simultaneously with the mining activities and continued for decades or even centuries after the mining ceased.

Placer Mining

Placer deposits are generally in or near mountainous regions. The placer metals eroded from bedrock outcrops are carried down into stream channels, but the metals are seldom transported more than a few tens of miles from the bedrock source. The metals disseminated through bedrock outcrops as lode deposits are emplaced by processes associated with mountain building. As molten rock is pushed upward from deep within the Earth's interior, it folds and fractures the overlying rocks, uplifting them into mountains. Superheated water is explosively released from the molten rock as it rises. This water carries some elements in solution, and the water dissolves more elements from the surrounding rock as it moves upward through fractures in the rock. As the water continues to rise, changes in temperature, pressure, and water chemistry eventually cause metal carbonate and sulfide compounds to be deposited along the fractures. These processes form a three-dimensional mosaic of crisscrossing veins of enriched metal within the rock matrix. Because of the association between placer and lode deposits, what began as a placer-mining district worked by individuals or small teams of miners often gave way to a lode-mining district dominated by commercial operations.[27]

Placer sediments may underlie the contemporary bed of a river, or they may be in deposits left by ancient rivers across a valley bottom or along valley side slopes. The great majority of gold recovered from stream placers was found in the lowermost three feet of gravel and in fractures and potholes in the uppermost foot of the bedrock. Thus the miners often removed the sediments from the streambed and banks down to the bedrock contact and back to the valley walls, dumping the waste sediment back into the stream once they had removed the fragments of valuable metals.[28]

Lode mining from bedrock outcrops requires several stages of processing to remove small amounts of metal dispersed through the bedrock. In contrast, placer mining generally requires fairly simple separation processes based on the greater specific weight of the metals relative to the surrounding sediment, or based on chemical affinities of

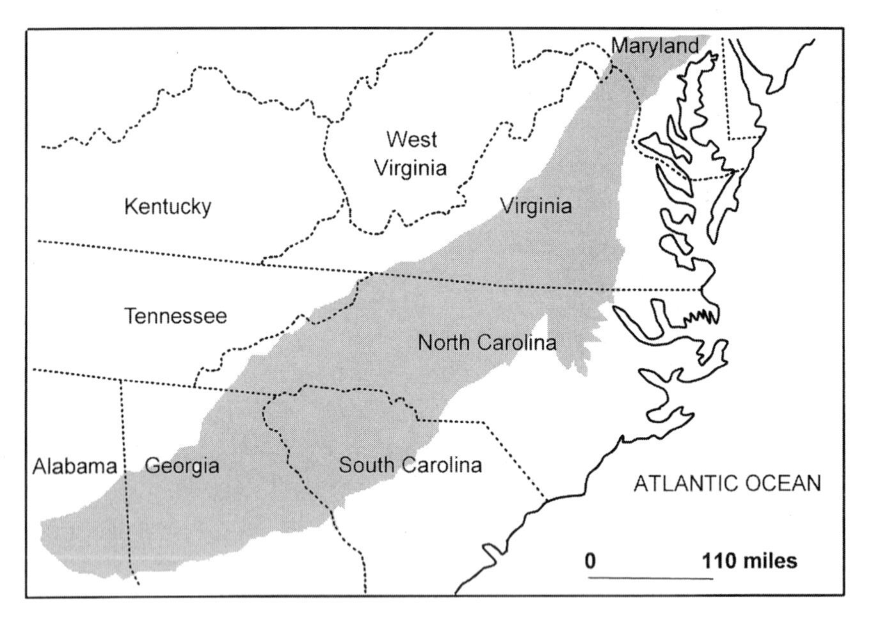

Shading indicates the principal gold belt of the southern Appalachians. (After H. B. C. Nitze and H. A. J. Wilkens, 1897, Gold mining in North Carolina and adjacent south Appalachian regions, *North Carolina Geological Survey Bulletin*, No. 10, Figure 1.)

the metals. Separation based on specific weight involves agitating the sediment-metal mix in water and then flushing off the lighter sediments by means of a handheld pan with ridged sides, or some type of rocker box or flume with a ridged bottom. The denser metals are trapped in the container's ridges and may then be removed. If mercury is added to the sediment-water mixture, the mercury adheres to the gold and sinks, making the gold easier to remove. Using either approach, an individual miner or a small team of miners can effectively work a placer deposit. A skilled miner with a gold pan can process about five-tenths to eight-tenths of a cubic yard of sediment in ten hours. Two miners operating a rocker box or hydraulic system, using between one hundred and eight hundred gallons of water, can process three to five cubic yards of sediment in the same ten hours.[29]

Many rivers have a coarse-grained surface layer of cobbles and boulders, with finer gravel and sand beneath. The wholesale disruption of the streambed during placer mining dramatically increased the ability of the water to carry the finer subsurface sediment downstream by disrupting the packing and frictional resistance of the sediments. As sediment

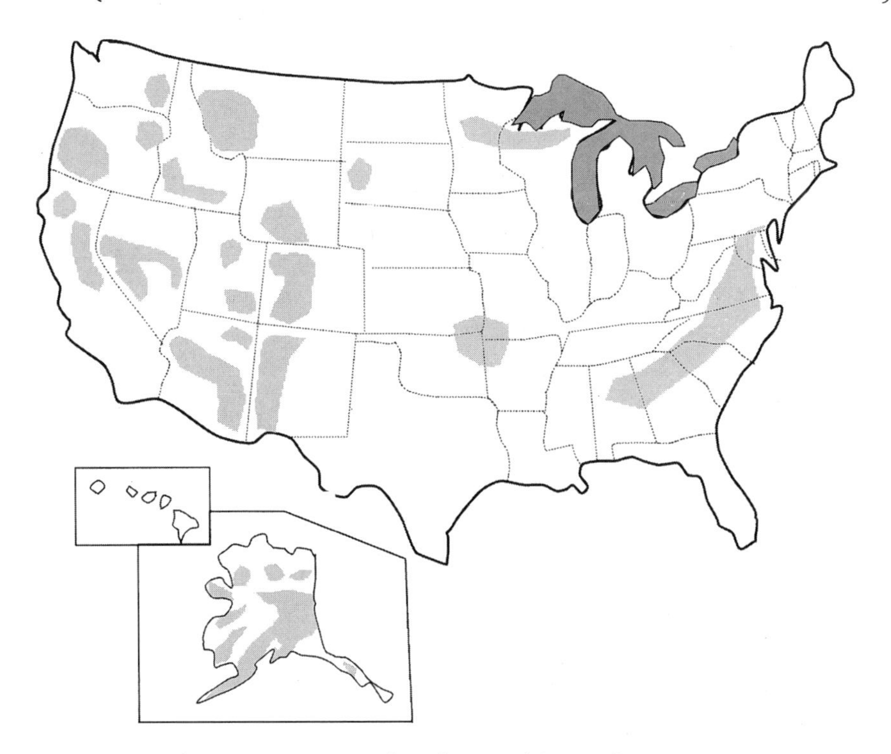

Primary areas of mining precious metals in the United States, 1800–2000.

transport increased, water quality dropped. Pools filled with sediment; whole channels filled with sediment and forced the stream flow over the banks; channels became unstable and moved back and forth across valley bottoms in braids; and stream plants and animals were stressed or eliminated.[30]

The earliest placer mining in the United States began in the southern Appalachians during the 1820s. This mining was much more localized and involved much smaller volumes of material than the mining that began in 1849 in California and spread throughout the western United States and to Canada, Australia, and Alaska during the succeeding decades. Although mining for precious metals occurred in the eastern United States and in Hawaii, metal mining was much more widespread and economically important in the western United States. The remainder of this chapter therefore focuses on the western region of the country.

Mining dominated the economy of various portions of the American West from the mid-1860s until the end of the twentieth century. It was

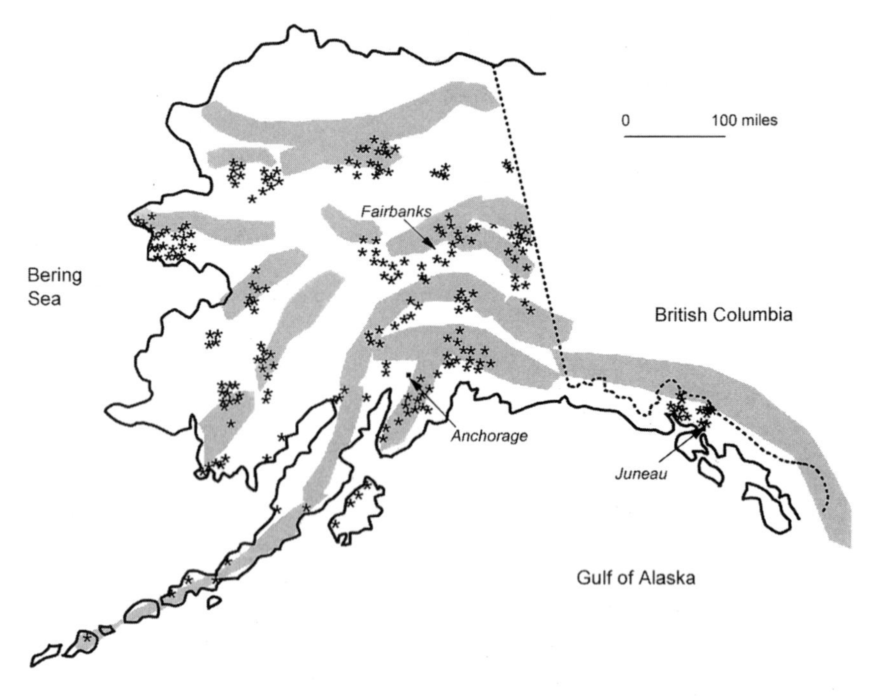

Principal placer mining districts of Alaska are indicated by asterisks. Mountain ranges are shaded. (After E. H. Cobb, 1973, Placer deposits of Alaska, *U.S. Geological Survey Bulletin* 1374, Figures 1 and 2.)

also vital to the national economy; the $785 million in gold mined from California by 1865 was crucial in financing the Civil War.[31]

The first period of mining at a particular site focused on nuggets and very coarse flakes of metal. But the site was often reworked repeatedly over many decades using techniques that extracted progressively finer particles of metal as the national economy and the prices of precious metals fluctuated. In California, a court decision in 1884 banned the discharge of mining sediment into streams and thus shut down hydraulic mining unless the sediment could be detained behind a dam. This mining resumed at a more limited scale in 1893, with a brief but widespread resurgence in the 1930s because the miners were able to build concrete arch dams that effectively contained the sediment. The Great Depression of the 1930s created new gold rushes in California, South Dakota, Colorado, Utah, Arizona, Idaho, Montana, and New Mexico.[32]

Most of the placer mining during the 1930s was much smaller in scope than mining during the nineteenth century. The U.S. Bureau of

Table 3.1. History and Extent of Major Placer Mining and Lode Mining Impacts to Streams in the United States

Location	Dates	Description
Southern Appalachians	1820s–1890s	Southwestern Virginia, West Virginia, eastern Kentucky and Tennessee, western North Carolina and South Carolina, northern Georgia, and northeastern Alabama: mining copper, gold, lead, zinc, coal, and salt back to prehistoric times. 1820s–49 gold lode mining in northern Georgia; 1828–50s, North and South Carolina. North Carolina: rocker boxes in use as early as 1825, hydraulicking after 1853, dredging in 1843 and continued through 1890s. Also used ditches and diversions. Most gold came from placers, but small volumes compared with western U.S. deposits. (See map.)
California	1849–1950s	Gold discovered at Sutter's Mill in central California in 1848. Hydraulicking invented ca. 1853 and operated on a large scale in the northern Sierra Nevada from the 1850s to the 1880s. More than 1.5 billion cubic yards of placer gravels worked. The Sawyer Decision in (1884) prohibited discharge of mining debris in Sierra Nevada region, but hydraulicking continued until 1950s in Klamath-Trinity Mountains. From the mid-1800s to the early 1900s most California gold came from lode and drift mining. After the early 1900s dredging of gold-bearing sediments in the Sierra Nevada foothills became an important source of gold. Mercury was used extensively until early 1960s dredging of floodplain deposits, and more than 3.6 billion cubic yards were mined this way. (See maps.)
Comstock lode, western Nevada	1850–80 (placer and lode)	Placer gold discovered in May 1850 at Gold Canyon, a small tributary of the Carson River near Reno. About one hundred people worked the deposit between 1852 and 1855, with desultory activity until the 1859 discovery of the Comstock lode of gold and silver. The California goldfield activity had been depressed in 1859 by the rush to the Fraser River placers in British Columbia, but the Comstock discovery reinvigorated California and the intermountain West (Colorado in 1860; Idaho, Montana, and Oregon in 1861), and lode mining soon

Table 3.1. Continued

Location	Dates	Description
		followed. The height of Comstock lode mining occurred between 1860 and 1880. In 1873 the miners diverted water from Sierra Nevada rivers in a huge pipe. The Sierras were devastated for about one hundred miles to provide the 600 million feet of lumber that went into the Comstock mines and the 2 million cords of firewood consumed by mines and mills up to 1880. Flumes many miles long were used to float timber down from the Sierras for many years; timber was also floated down the Carson River.
Colorado	1859–1950s	Central City in Colorado Front Range: placer, 1858; lode, 1859 to early 1900s—gold, silver, copper, lead, and zinc. San Juan Mountains lode, 1877 to present. Alma in South Park, placers and lode, 1860 to present—gold, silver, lead, zinc, and copper. South Platte River basin placers, 1870s to 1942 and later. Leadville gold and silver placers to present with lode mining.
Southern Idaho	Early 1860s–1930s	Placer gold in Snake River, primarily between King Hill and American Falls; hydraulic mining.
Oregon	1860s–1940s	Placer gold; hydraulic, dredge, sluice, and hand operations. Primarily southwestern and northeastern Oregon.
Montana	1860s–present	Butte region; 1864 gold, 1867 silver, 1879 copper. The Clark Fork River has received excess heavy metals since the beginning of lode mining in the Butte area in the 1860s and on-site milling of metals in 1883.
	1890s–1960s	Upper Blackfoot River drainage, west-central Montana: placer gold discovered in 1852. Silver and lead mines in upper Blackfoot most active from 1890s to 1960s. Acid-mine drainage widespread. Tailings dam failure in 1975.
Utah	1868–present	Park City lode mining district discovered ca. 1868. Through 1982, produced 1.45 million ounces of gold, 253 million ounces of silver, 2.7 billion pounds of lead, 1.5 billion pounds of zinc, and 128 million pounds of copper. Tintic lode mining district, 1869–1990s.

Table 3.1. Continued

Location	Dates	Description
Black Hills, South Dakota	1875–90s	Gold rush began with 1875 discovery of placers at Deadwood. Gold and silver deposits in Black Hills have produced more than 3 million troy ounces, excluding the Homestake Mine.
Alaska	1893–present	Placer mining begun as early as 1834 by Russians, but only small-scale pans and rocker boxes. Placer strike in 1893 on Koyukuk River mostly finished by early 1900s (miners moved to 1902 Fairbanks placer strike). Discovery of gold placer in 1896 on a tributary of the Klondike River in the Yukon Territory of Canada set off the 1897–98 Klondike Gold Rush. Placers began to be worked with hydraulic systems and dredge boats. Gold eventually found to be widespread below the Arctic Circle; 1880 Juneau, 1884 Forty Mile River, 1893 Birch Creek in Circle Mining District, 1898 Nome, 1902 Tanana River and Fairbanks, 1906 Nolan Creek. Mining continued in some region until the 1920s, then revived in the 1930s and 1970s and 1980s; some areas still active today. Of all the gold mined in the world, 1 to 2 percent (33 million ounces) has come from Alaska, and 72 percent of this has come from placer deposits. Mining continues to be substantial. In 1993, for example, gold was mined at 196 placer mines. (See map.)

Sources: G. H. Smith, 1998, *The history of the Comstock lode,* University of Nevada Press, Reno, Nev.; D. Stiller, 2000, *Wounding the West: Montana, mining, and the environment,* University of Nebraska Press, Lincoln; R. Manning, 1997, *One round river: The curse of gold and the fight for the Big Blackfoot,* Henry Holt and Co., New York; H. B. C. Nitze and H. A. J. Wilkens, 1897, Gold mining in North Carolina and adjacent south Appalachian regions, *North Carolina Geological Survey Bulletin,* No. 10; W. Yeend, P. H. Stauffer, and J. W. Hendley, 1998, Rivers of gold—placer mining in Alaska, *U.S. Geological Survey Fact Sheet 058-98;* J. J. Norton, C. S. Bromfield, D. R. Shawe, T. B. Nolan, and A. R. Wallace, 1989, Gold-bearing polymetallic veins and replacement deposits—Part I, *U.S. Geological Survey Bulletin 1857-C;* H. T. Morris, F. S. Fisher, D. R. Shawe, and T. B. Thompson, 1990, Gold-bearing polymetallic veins and replacement deposits—Part II, *U.S. Geological Survey Bulletin 1857-F;* K, Capps and P. Tacquard, 1999, The search for gold along the Koyukuk River, *Bureau of Land Management Report BLM/AK/GI-99;* S. L. Yarnell, 1998, The Appalachians: A history of the landscape, *USDA Forest Service General Technical Report SRS-18;* L. Ramp, 1960, Gold placer mining in southwestern Oregon, *The Ore-Bin,* 22, 75–79; R. E. Rohe, 1983, Man as a geomorphic agent: Hydraulic mining in the American West, *Pacific Historian,* 27, 5–16; R. E. Rohe, 1984, Gold mining landscapes of the West, *California Geology,* 37, 224–30; R. E. Rohe, 1984, Just scratching the surface: Geographers and the mining West, *Geographical Bulletin,* 26, 35–46; R. E. Rohe, 1985, Hydraulicking in the American West: The development and diffusion of a mining technique, *Montana,* 35, 18–34.

Mines and the various state mine agencies fostered the 1930s rushes by printing "how to mine" manuals. Between fifteen thousand and twenty-five thousand people worked small-scale placers with pans, rocker boxes, mercury, and sometimes dredging. Most of the rushes began around 1931 and peaked in 1934. The Roosevelt administration's guarantee of gold and silver prices in 1934 caused a resurgence in commercial lode mining and placer dredging, and by 1937 the rush of individual miners was largely over. Placer mining continues today in many parts of the United States but is much less widespread and intensive than it was historically, with the exception of contemporary mining in Alaska.[33]

Legal regulation of mining lagged behind discoveries of precious metals, and the first miners largely made their own rules. Mining camp law ranked claims according to age, subject to continuous occupation and use. This priority system was important because of the volumes of water necessary to work a placer claim. A downstream claim with sizable capital investments might become effectively unworkable if a subsequent upstream claimant diverted too much water from the stream channel.[34]

The placer miners in California argued for several years over whether water use should be governed by prior appropriation or riparian rights. Prior appropriation allocated water based on date of claim ("first in time, first in right"); riparian rights allocated water based on proximity to the stream channel. Many of the miners in the western regions initially favored riparian rights. The miners argued that the hard-working individual prospector was better served by riparian rights, whereas prior appropriation catered to the demands of investors in corporate ventures. These corporations had to rely on the water guaranteed via prior appropriation to justify initial capital expenditures. In 1854 the Ninth Judicial District Court for Trinity County, California, which contained the mining district of Weaverville, upheld prior appropriation. By that time, prior appropriation served the interests of the largest number of miners on the stream. The water-intensive hydraulic mining that became more widespread by 1859–60 was particularly dependent on prior appropriation.[35]

Western courts tended to confirm the rights of private corporations after small-scale operations had dramatically declined in a particular district. Prior appropriation was never complete, for mining districts remained isolated and autonomous, and some riparian rights remained. But prior appropriation gradually came to dominate all forms of water use in the western United States. As Donald Pisani wrote in his history of water, land, and law in the western United States: "this new system of water law [prior appropriation] was remarkably consistent with the American ideal of limited government. Prior appropriation was one of the greatest nineteenth-century legal subsidies in that it allowed public property (water on public land) to be taken for free."[36] And once taken, those water claims became invested with almost sacred legal rights that have prevented any subsequent serious reevaluation of water allocation systems in the western United States.

The first federal mining law, in 1866, was designed largely to protect investors in lode mining. This law effectively opened the American West to mining, and it was followed in 1870 by a law covering placer deposits. Both laws allowed free and open access to public lands. The General Mining Law of 1872 combined and codified the two earlier laws, limiting placer claims to twenty acres in extent and requiring at least one hundred dollars worth of work annually to maintain a claim. The law recognized the property rights schemes of existing mining operations. It also allowed a miner to buy or patent land for five dollars an acre and to pay no royalty to the federal treasury for minerals extracted. Historians interpret the 1872 law as reflecting the lack of a national administrative state, as well as the contemporary policies of economic liberalism and reduced governmental intervention.[37]

All mining claims are now recorded with the Bureau of Land Management, but three agencies at the Department of the Interior are involved in regulating mining: the Bureau of Land Management, the Minerals Management Service, and the Office of Surface Mining. Individual states can also regulate lode mining. Although the Clean Air Act, the Clean Water Act, Superfund legislation, and other federal policies enacted during the 1970s and 1980s affected mining, the 1872 mining law remains the national law governing the filing and maintenance of mining claims on public lands.[38]

Congressional representatives from western states have blocked repeated attempts to reform the present complicated and antiquated system of regulating mining in the United States. Besides robbing the public and the federal government of valuable royalties by allowing mineral rights to be claimed at costs far below market value, the 1872 law helps to encourage environmental irresponsibility. When mineral claims are cheap, prospecting is more likely to be extensive and wasteful.[39]

The physical impacts to rivers of activities associated with placer mining occur both in the river and on the adjacent hillslopes. As mentioned previously, the working of sediments in and near river channels greatly increases the mobility of the sediments, leading to increased turbidity in the river and increased sediment deposition downstream. One of the earliest published comments on the potentially destructive effects of excess sediment in streams was David Starr Jordan's 1889 remarks on the loss of trout habitat in Colorado streams affected by mining. However, as late as 1938 the *California Mining Journal* claimed that "young fish thrive on mud."[40]

Placer gold mining increased substantially in Alaska during the 1970s after the deregulation of gold prices. Most contemporary understanding of the immediate, direct impacts of placer mining on stream ecosystems comes from studies conducted during the 1980s and 1990s along subarctic streams in Alaska on which placer gold mining is still occurring. These studies found that algal production, the base of many river food webs, is reduced by half along moderately mined streams relative to unmined streams. Placer mining sediment affects bottom algal production both by limiting light penetration and by smothering and scouring the algae. Algal production is undetectable in the turbid waters of heavily mined streams. Fine sediment deposited on streambeds fills the spaces between gravels and reduces gas exchange between the surface and subsurface water. This reduction in dissolved oxygen is harmful to the fish eggs and larvae and the invertebrates ordinarily found among the streambed sediments. Sediment from placer mining is associated with decreased density and biomass of invertebrates. Fish suffer impaired feeding activity because of reduced sight and decline in numbers of prey. The fish also experience gill abrasion, reduced growth rates, downstream displacement, decreased resistance to environmental stresses, and in cases of extreme sediment loading, death. The indirect effects of sedimentation, through loss of shallow-water, summer habitat for feeding and reproduction, more severely affect fish populations than the direct effects of sedimentation affect the health and survival of individual fish.[41]

Increased downstream sediment deposition from mining alters the grain-size distribution of the streambed and the volume of pools, thus changing aquatic habitat. Studies of Alaskan rivers with a wide range of water depths and habitats indicate that after mining these rivers have few deep pools and little habitat diversity. Habitat recovery requires many large floods and is predicted to take more than a decade on these streams. Recovery from mining requires many decades along some rivers. Deposition of mining sediment can so destabilize a stream that it spills over its banks more frequently during higher flows or begins to move laterally across the valley bottom. After the start of placer mining on California's Yuba River in 1849, the river bed rose twenty feet in elevation at Marysville, more than twenty miles downstream from most of the mining camps. The lingering effects of nineteenth-century mining continue to impact streambed elevations and stability along the Yuba River during the twenty-first century.[42]

Diversion of water to work a placer claim affects both the source stream and the receiving stream by changing the volume and timing of stream flow, as well as the water temperature and chemistry. These changes affect aquatic and riverside organisms whose life cycles are governed by the natural flow regime.[43]

Contaminants other than sediment can be introduced to rivers as a result of mining activity. Chemical contaminants come primarily from mercury used in the amalgamation process or from acid-mine drainage of lode mines and tailing piles. Miners used mercury in placer and lode mines throughout the United States, with hydraulic mining of placer deposits producing the most environmental contamination. In a typical sluice operation, miners added hundreds of pounds of liquid mercury to their riffles and troughs to enhance gold recovery, pouring the mercury from seventy-six-pound barrels. The density of mercury is between the density of gold and the density of a gravel slurry, so the gold and mercury-gold amalgam sink. However, many fine gold and mercury particles are washed through and out of the sluice, even when an undercurrent (a lower, perpendicular flume into which fine sediment drops) is used. Historical accounts describe minute mercury particles found floating on the water surface up to twenty miles downstream from a sluice.[44]

The average placer operation in California used one- to four-tenths of a pound of mercury per square foot of sluice. A typical sluice had twenty-four hundred square feet of area and used up to eight hundred pounds of mercury initially, with later additions as required. From 10 to 30 percent of the mercury was lost downstream during amalgamation, so even a few placer operations could annually add several hundred pounds of mercury to a river. A century after mining ceased, each of the many placer sites in California still has hundreds to thousands of pounds of mercury dispersed in the adjacent soils and waters, as well as elevated mercury levels in invertebrates, amphibians, and fish downstream. U.S. Geological Survey scientists examining the California sites in 2000 estimated that the hundreds of sluices operated between the 1860s and the early 1900s used twenty-six million pounds of mercury, of which at least three to eight million pounds, and probably much more, were lost into the environment. A strong regional correlation remains between mercury bioaccumulation in stream organisms and the intensity of hydraulic mining in the Sierra Nevada. The highest average levels of mercury bioaccumulation occur in the intensively mined Bear and South Yuba watersheds.[45]

Some of the dredge tailings sites along California's American River were leveled and used for housing developments. In addition to the possibility of surface and groundwater contamination by mercury in these developments, elemental mercury and toxic mercury compounds may be transformed directly from a solid to a gaseous state by application of organic matter on lawns. This effect can be prevented by the application of elemental sulfur, which binds to mercury and creates an insoluble, nontoxic compound, but the developers or homeowners must be aware of the mercury contaminants.[46]

Mercury is nasty stuff. It and its compounds have no known biological function, and they interfere with most biological functions. Most concentrations of mercury come from human activities, including agriculture, mining, and manufacturing. Elevated levels of mercury within organisms living in mercury-contaminated areas may persist for a century after the source of pollution is discontinued. Forms of mercury with relatively low toxicity can be altered to forms of very high toxicity through biological processes. Some forms of bacteria that thrive in the conditions of low dissolved oxygen present in sediment or in algal mats transform oxidized mercury to methylmercury. This methylmercury is then taken up by other organisms. The difference between tolerable natural background levels of mercury and harmful effects in the environment is exceptionally small. Mercury can be lethal to sensitive aquatic organisms at only one-tenth to 2 parts per billion in water, at 2.2 parts per million body weight for birds, and at one-tenth parts per million body weight for mammals. These nearly undetectable, tiny amounts of mercury can be bioconcentrated within an organism and biomagnified through a food web. As the mercury accumulates in an organism's body, it interferes with the normal functioning of cells, resulting in mutations, developmental abnormalities, cancer, or death.[47]

As with other legacies from the pioneer era, we will live with mining's toxic wastes for decades or even centuries to come. Acid-mine drainage of surface and groundwater with high concentrations of contaminants from abandoned lode mines or mine tailings can release a variety of toxic materials to the environment, depending on the metals present in the mine and the methods used to process them. Metals that may be toxic to living organisms when released from mine sites include aluminum, antimony, arsenic, calcium, copper, chromium, cadmium, magnesium, mercury, nickel, lead, selenium, silver, and zinc. The broad array of metals released from the point site of a mine disperses through the environment like an epidemic disease. Arsenic, for example, attaches to clay particles and organic matter, gradually disseminating downstream, across the floodplain, and into aquatic and riverside organisms. As of 2000, more than 557,000 abandoned mines in thirty-two states—most of them in the West—accounted for fifty billion tons of untreated mine waste polluting twelve thousand miles of waterways and 180,000 acres of lakes and reservoirs. Any living organism, from an insect to a human, that comes into contact with these thousands of miles of poisoned rivers and lakes can ingest and store these toxins in its own body. Arsenic in the drinking-water wells of Fairbanks, Alaska, is associated with dredging for placer gold before World War II. Plants along Soda Butte Creek in Yellowstone National Park are less diverse and abundant fifty years after a flood deposited trace metals eroded from mine tailings upstream and beyond the park boundary. Mercury levels in river and floodplain sediments along the Chestatee and Etowah River systems of Georgia exceed state and federal standards half a century after gold mining in the region ceased.[48]

Some of the more grotesque examples of mine-generated contamination come from Montana. In June 1975 the Mike Horse Mine dam failed during a flood caused by heavy rainfall and rapid snowmelt. The sixty-foot-high, 450-foot-wide dam impounded thirty years' worth of mine tailings. When the dam failed, two hundred thousand cubic yards of tailings flushed ten miles downstream into Beartrap Creek and the Blackfoot River. Over the succeeding years, the tailings moved farther downstream. The numbers of fish and bottom-dwelling stream organisms dropped by 65 to 85 percent in the affected reaches of river. The diversity of aquatic organisms also dropped. Several years later, scientists from the University of Montana found abnormally high concentrations of aluminum, cadmium, copper, iron, lead, zinc, and arsenic in sediment and in the tissues of various organisms. They also found pronounced mobility of the metals fifteen miles below the mine and far beyond the wetlands that were presumed to act as buffers to metal and sediment transport. The scientists found decreased trout populations and trout and stone flies that had concentrated zinc in their bodies. These organisms were present as far as forty-six miles downstream in the 1990s, and the effects will presumably spread farther in the future as the metal-contaminated sediments work their way downstream. An ecologist from Montana's Department of Fish and Game compared tailings impoundments to "time bombs that are scattered wherever there has been mining, and which are just sitting there waiting for circumstances to cause something like this [1975 dam failure]."[49]

The Clark Fork River of western Montana is another startling example of mining-generated contamination, with the Butte copper mining area as the source of the contaminants. The nine-hundred-foot-deep waters of the Berkeley Pit in Butte kill geese unfortunate enough to land there. Contaminated groundwater seeping out of Butte enters Silver Bow Creek, as does surface runoff contaminated by mine tailings. Silver Bow Creek flows into the Clark Fork River, taking its wastes with it. Heavy metals permeate the streambed and floodplain sediments along the Clark Fork. Since 1908 a small reservoir at Milltown, 150 miles downstream from Butte, has trapped metal-laden sediments. The 140 miles of the Clark Fork between Butte and Milltown are toxic enough to kill fish. Arsenic moving from the sediments of the Milltown Reservoir has further contaminated local groundwater, and arsenic in the surface sediment is six times the background levels. In the older sediments at depth, arsenic is thirty-two times background levels. Zinc and copper are sixty times background levels.[50]

Arsenic is a relatively common element that is nutritionally essential or beneficial, but at higher levels it acts as a carcinogen and a teratogen, which produces developmental abnormalities. Arsenic can be absorbed by an organism through ingestion, inhalation, or permeation of skin or mucous membranes; and once in an organism, it can be bioconcentrated. Sensitive aquatic species are damaged at water concentrations of 19 to 48 parts per billion, 120 parts per million in the diet, or tissue residues greater than 1.3 parts per million in freshwater fish.[51]

Higher levels of zinc are also detrimental to aquatic life, although zinc is essential to living organisms in small amounts. Zinc primarily affects zinc-dependent enzymes that regulate the activities of RNA and DNA in living organisms. Zinc tends to accumulate in the gill tissue of fish and in the pancreas and bones of birds and mammals. Zinc interacts with many other chemicals, sometimes thereby increasing in toxicity. It mostly accumulates in sediments, although it can bioaccumulate in some organisms. Zinc is most harmful to aquatic life under conditions of low pH and dissolved oxygen as well as warm temperatures. The most sensitive aquatic species are affected at concentrations as low as ten to twenty-five parts zinc per billion parts of water. Typical background concentrations of zinc are less than forty parts per billion in water and two hundred parts per million in soils and sediments. At high concentrations zinc causes developmental changes and abnormalities, or death.[52]

Like zinc, copper is one of those elements for which a small amount is a good thing but a large amount is deadly. Copper is naturally plentiful and essential for the normal growth and metabolism of all living organisms. However, it is also among the most toxic of heavy metals in freshwater organisms. Copper often accumulates and causes irreversible harm to some species at concentrations just above the levels required for growth and reproduction. Uncontaminated rivers have copper concentration levels of one to seven parts per billion, whereas contaminated rivers may reach or exceed fifty to one hundred parts per billion. Copper can induce cancerous growths, chromosomal changes, developmental changes and abnormalities, and death at high concentrations. It also interacts with numerous organic and inorganic compounds, resulting in altered bioavailability and toxicity. Copper disrupts the gill tissue of invertebrates. In fish, copper interferes with the processes by which cells regulate their biochemistry and leads to death from tissue destruction. Studies of the effects of copper on microorganisms in streams of California's Sierra Nevada found that copper inhibits in-stream photosynthesis, one of the bases of the river food web. The species composition of the bottom-dwelling algae shift to an assemblage more tolerant of copper. Copper also inhibits the rate of processing of leaf litter by microorganisms, the other base of the river food web.[53]

The zinc, copper, and arsenic leaking from the Milltown dam led to its designation as a Superfund site. The 1980 Comprehensive Environmental Response, Compensation, and Liability Act (CERCLA), the Superfund act, requires responsible parties to clean up environmental contamination. At sites where the responsible parties are long gone, as in many of the nineteenth-century mining regions, remediation costs are paid from a tax on the chemical industries that supplies the Superfund. The Superfund program had some notable successes in containing and eliminating sources of toxic contamination until its funding was cut by the Bush administration in 2002.[54]

Effective management of mining-contaminated sediments is costly and difficult, requiring identification, regulation, and remediation. Given more sites than can be immediately treated, identification is not necessarily straightforward. A triage approach is necessary to determine the most dangerous sites. There are questions of what medium to test. Is contamination best reflected in sediments, in water, or in living organisms? Should determination of contamination be based on universal standards, or on site-specific criteria? Should we use standards for individual chemicals, or evaluate multichemical mixtures?[55]

Regulation of mining-contaminated sediments is equally uncertain. As one writer remarked, "Given the pervasiveness of contaminated sediments and their potential environmental and economic impacts, it is remarkable that the legal and legislative communities have generated so little commentary on the regulation and remediation of contaminated aquatic sediments."[56] At the start of the twenty-first century, the United States has little legal precedent for contaminant regulation. Few have attempted to address who has the authority to develop contaminant criteria and evaluative protocols, what sediments should be covered by standards, or how standards may be enforced. The Army Corps of Engineers and Environmental Protection Agency (EPA) at the federal level, and various state agencies at the local level, undertake evaluation and enforcement based on legislation including the National Environmental Policy Act (1970), the Federal Water Pollution Control Act (the Clean Water Act, 1977), CERCLA (1980), and EPA drinking water standards, but oversight remains inconsistent and often ineffective.

Remediation of mining contamination may involve no action if natural sedimentation is likely to bury and contain the contaminants, or if natural degradation and solution processes can reduce contaminant loads. If this is not possible, in-place containment with capping or lateral confinement and/or treatment that solidifies the sediments or immobilizes the contaminants may be used. Remediation may also involve dredge removal and disposal of contaminated sediments from sites where environmental impacts are severe. Removal and disposal may be necessary where flooding or river erosion can mobilize sediments, or where navigation ways must be dredged. Disposal generally occurs at confined sites. It may include treatment that removes the fine sediments carrying contaminants and thus reduces waste volume. Disposal may also include treatment that immobilizes the pollutants, extracts and recycles them, or destroys them. The various types of treatment ranged in cost from 45 dollars to 750 dollars per cubic yard in the late 1980s. The 1995 reclamation of three thousand feet of stream along Whites Gulch, a historically placer mined valley in southwestern Montana, cost more than half a million dollars. Our 1872 decision to maximize profit to individual miners and mining companies continues to cost society dearly—and to include hidden costs of endangered human and ecosystem health that we are not even able to quantify.[57]

Where placer mining is still occurring, as in Alaska, downstream impacts can be reduced by stipulating guidelines for mining wastes. Diversion of surface waters away from the mining site reduces downstream transport of waste. Construction and maintenance of water retention structures allows settling of sediment and contaminants. The miners can be required to provide assurance that removed pollutant materials will be retained in storage areas such as settling ponds. And new water can be limited to the minimum amount required for processing operations. Alaskan state regulations stipulate that suction dredges may not increase the turbidity of the river by more than five turbidity units at five hundred feet downstream, and settling ponds are also required. On a broader scale, recommendations to reduce mining impacts include saving and ultimately respreading topsoil; controlling toxic substances and sediment discharge; continuing stabilization of mined areas in and along streams; making provisions for fish, wildlife, and human passage during mining; and rehabilitating habitat and vegetation after mining. In Alaska, the Bureau of Land Management (BLM) oversees this process by the use of performance standards. These standards are partially subjective guidelines that are formulated on the basis of existing legal requirements together with site-specific judgments by BLM employees.[58]

Ultimately, the effectiveness of any guidelines and the actions that such guidelines stipulate rests on the willingness of individual miners and corporations to follow the guidelines, and on the ability of responsible agencies such as the BLM to enforce the guidelines. Simply put, environmental protection from mining contaminants depends on attitude and budget.

People also try to restore rivers disturbed by placer mining. The National Park Service restored portions of lower Glen Creek in Denali National Park and Preserve that had been placer mined from 1906 to 1941, and again in the 1970s. Restoration efforts began with the estimation of flow conditions, channel slope, and channel sinuosity from regional conditions. Once a range of channel configurations was determined, hydraulic equations were applied to each configuration to determine stability for the channel bed and banks. Floodplain design was based on estimates of the hundred-year flood flow. The channel and floodplain were recontoured using heavy equipment, and natural revegetation of the floodplain and channel banks was enhanced with alder and willow brush bars. These brush bars are partially buried bundles of cut willow and alder. They helped to dissipate flow energy and encourage sediment deposition during a moderate flood near the end of the two-year construction project, but the newly configured channel experienced some bank erosion and changes in bed slope and channel sinuosity. It remains unclear how successfully we can restore disturbed streams.[59]

Changes in watershed characteristics caused by deforestation, increased frequency and extent of wildfires, and construction of roads, railroads, and cities further affect rivers in mining areas. Tremendous amounts of lumber are used during placer and lode mining to build sluices and support tunnels in lode mines. Lumber also goes to build housing, stamp mills, and smelters, as well as railroads to mining districts. During the nineteenth century, wood provided fuel for heating, cooking, and smelting. Deforestation, and the wildfires often associated with it, tend to increase the movement of water and sediment from hillsides. This leads to flooding in rivers and, if the hillslopes are sufficiently destabilized to give way in debris flows and landslides, the rivers may be choked with sediment. The construction of roads, railroads, and cities also increases water and sediment movement from areas adjacent to river channels. Together, these changes further alter rivers already directly impacted by in-channel mining or acid-mine drainage.[60]

Mining in California

California is the poster child of placer gold mining. The discovery of gold in the American River at Sutter's Mill during 1848 set off North America's first great gold rush. The California miners pioneered mining techniques and legal frameworks that were then exported to other mining regions. The people who suffered the secondary effects of the sediment mobilized by mining—those with agricultural interests in the Sacramento River valley and commercial interests in San Francisco Bay—pioneered the legal framework that restricted or halted mining. The environmental consequences of the mining persist more than a century after the mining ceased.

The magnitude of the Sierra Nevada placer mining is difficult to grasp. Placer mining continued on a large scale for nearly forty years in the Sierra Nevada after the Sutter discovery, and those few decades of activity created havoc. Until shut down by a court ruling in 1884, hydraulic mining dislocated a quantity of sediment approximately equal to eight times that moved in excavating the Panama Canal. The more than one billion cubic yards of mining sediment produced throughout the northern Sierra Nevada between 1853 and 1884 caused widespread environmental change. The 1893 Caminetti Act legalized hydraulic mining if the sediment was detained, and the ensuing period of licensed mining from 1893 to 1953 produced at least an additional twenty-four million cubic yards of mining sediment.[61]

The primary mining region covered the principal tributaries of the Sacramento River, stretching more than two hundred miles south from the Feather River to the San Joaquin River and more than fifty miles west from the crest of the Sierra Nevada to their foothills. This area of approximately fourteen thousand square miles was the site of more than sixty primary mining camps and thousands of smaller diggings and the temporary home of hundreds of thousands of miners.[62]

Today, place-names such as Placerville and Placer County, and the advertising logos used by businesses, are the most obvious remains of the mining frenzy. At places such as China Flat on the Silver Fork, only

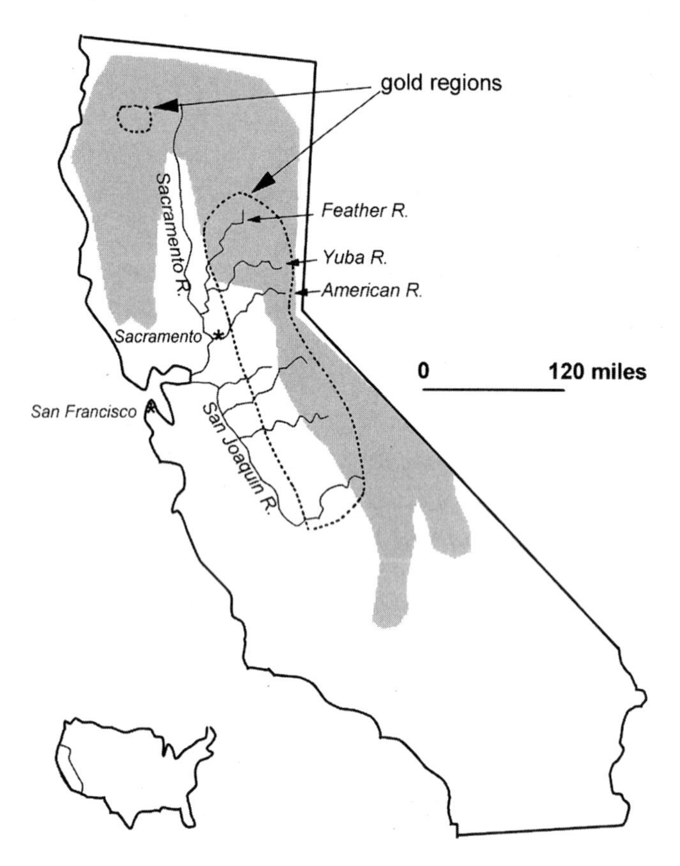

Location of primary placer mining districts and rivers in north-central California. Mountain ranges are shaded. Dashed line indicates the extent of the principal gold-mining region. (After M. J. Rohrbough, 1997, *Days of gold: The California gold rush and the American nation,* University of California Press, Berkeley, and D. Goodman, 1994, *Gold seeking: Victoria and California in the 1850s,* Stanford University Press, Stanford, Calif.)

old foundations remain as evidence of the previous activity there. The stream flows clear over brown and gray cobbles, and the air is fragrant with pine resin. There are large gravel bars downstream at the confluence of the North and Middle forks of the American River, but these are not unusual along bedrock canyons, and white-water boaters happily use the site. It is only farther downstream, at stretches such as the Yuba River near Marysville, where persistent accumulations of mining sediment give the rivers an air of desolation. For the most part, the dramatic topography of the Sierra Nevada and elevation-related changes in vegetation dominate the impressions of a casual observer.

The eastern Sierras rise abruptly from the sage scrub of the Great Basin deserts to steep slopes with open pine woodlands. Across the crest of the mountains the land drops more gradually toward the Pacific coast, but the vegetation changes dramatically. On the western side of the range the conifers are much larger, with moss clinging to their trunks and lush undergrowth. Outcrops of gray granitic rocks show here and there along the thickly wooded slopes of the steep, narrow river canyons. As the land drops steadily to the west, the pines give way to dense oak and manzanita chaparral, and the soil takes on a hue of burnt orange. Roads snake back and forth over the steep hills. It is difficult to imagine how miners ever found mineral deposits among the nearly impenetrable manzanita scrub. Thousands of prospectors swarmed across the rugged landscape like ants, always looking.

Many forces shaped the rivers of the Sierra Nevada before European Americans found gold along their channels. The mountain range is a huge mass of granitic rock intruded into the older, overlying rocks between approximately 150 and 160 million years ago. Along with the granitic magma came gold emplaced in veins as the magma hardened into rock. A long period of weathering and erosion removed the overlying rocks to expose the granitic core of the mountains, but during much of recent geologic history the core remained at two thousand feet above sea level. A major drainage system with gold-bearing gravels developed across the granitic core, only to be buried about 20 million years ago by a period of volcanic eruptions. Approximately 5 million years ago, forces in the Earth began to lift the core of the range. The uplift occurred in stages that continued until less than 2 million years ago, creating a giant wedge that towers ten thousand feet above the elevation of the adjacent Owens Valley. This wedge is the Sierra Nevada, a mountain range of stepped topography tilted toward the west.[63]

As the range was uplifted, new rivers cut deep canyons into the rock. The upper portions of the valleys that rivers began to carve into the tough granitic rocks were further deepened by the masses of glacial ice that repeatedly moved down from ice fields on the mountain crest during the past two million years. The glaciers scooped out broad troughs and carved giant steps into the valley bottoms. The tributary valleys were left hanging far up the main valley walls, so that when the ice melted the main valleys were strung with waterfalls along their sides. The glaciers carried tons of sediment with them, depositing it in linear mounds

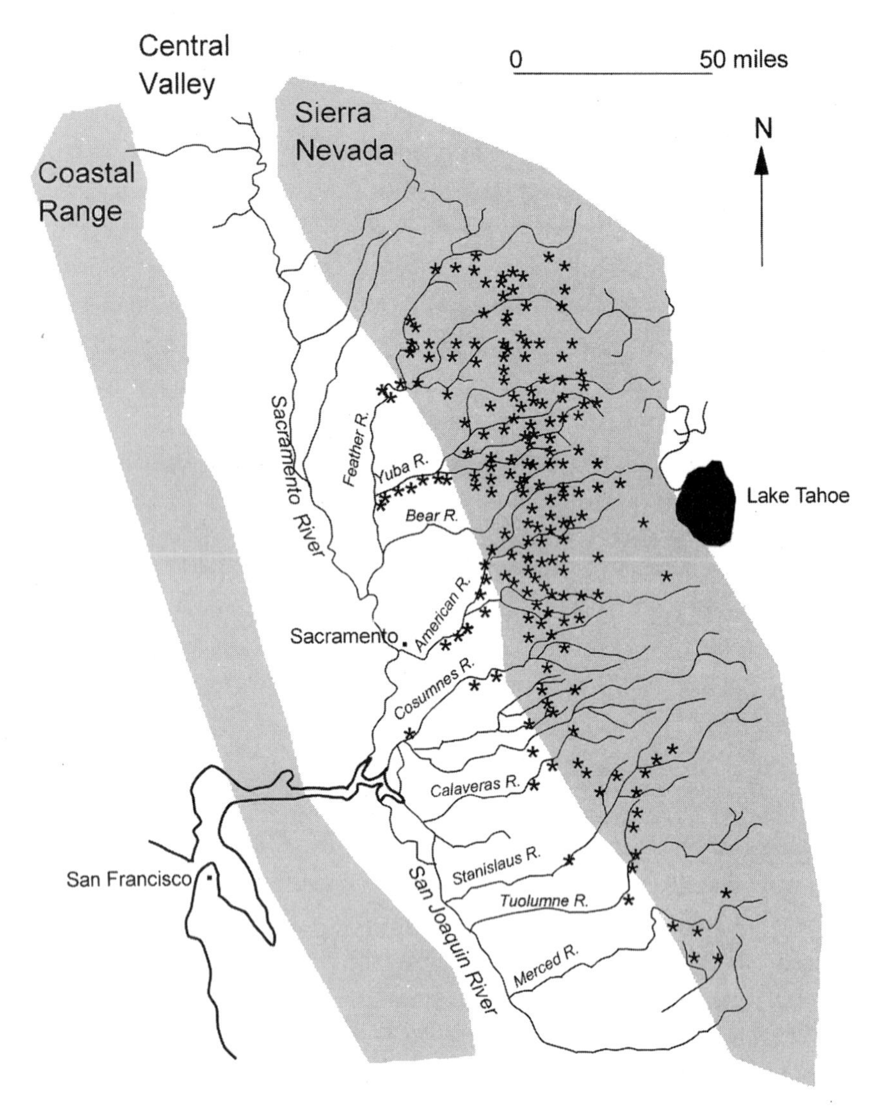

Left: Detail map of primary placer mining sites in the Sierra Nevada; mountains are shaded. (After C. N. Alpers and M. P. Hunerlach, 2000, Mercury contamination from historic gold mining in California, *U.S. Geological Survey Fact Sheet FS-061-00,* Figure 7; J. S. Holliday, 1999, *Rush for riches: Gold fever and the making of California,* Oakland Museum of California and University of California Press, Berkeley; E. G. Gudde, 1975, California gold camps, University of California Press, Berkeley.) Right: Detail of operations at a single placer site, in this case the Malakoff Mine along the South Yuba River. Shaded areas are hydraulic mining sites; dashed line indicates the top edge of a steeply sloping area. (After Holliday, 1999.)

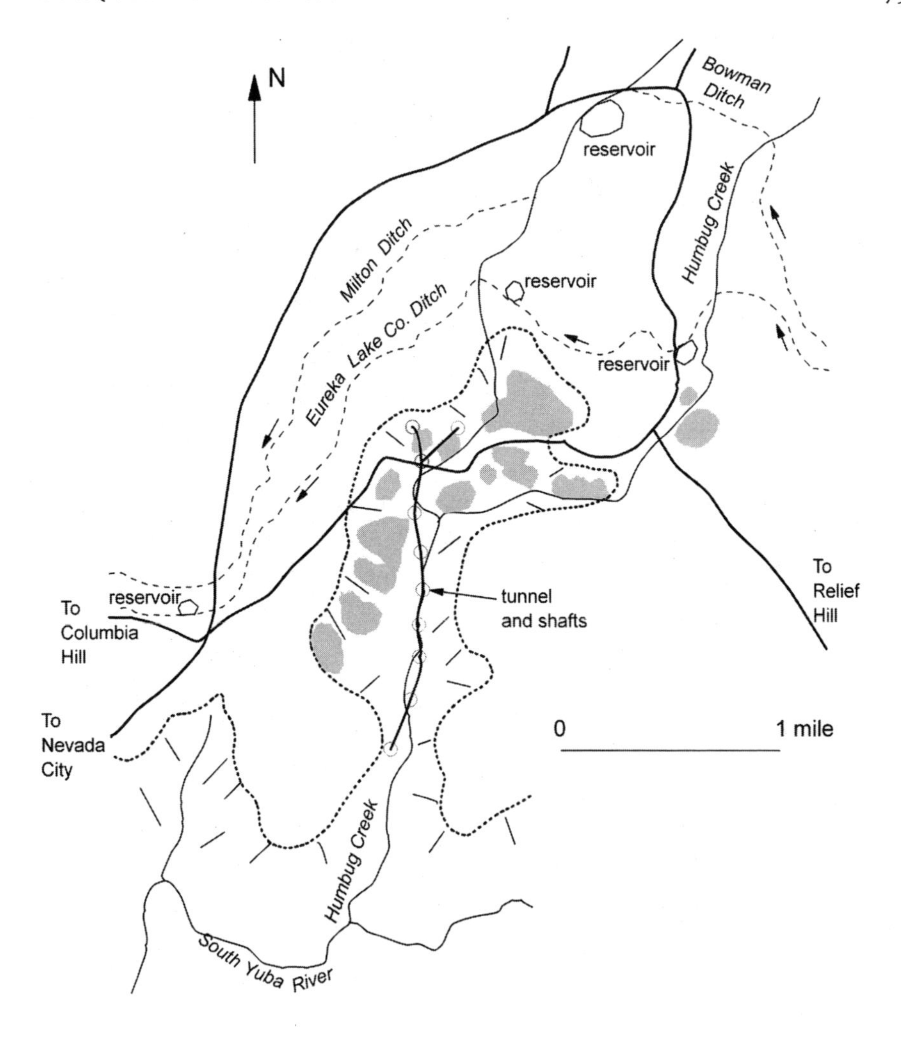

along the valley sides or across the valley at the terminus of the ice. Dispersed throughout this sediment were the nuggets and flakes of gold eroded from veins in the bedrock.

Once the ice was gone, the rivers began to gradually rework the glacial sediment and create their own courses along the altered upper valleys, while continuing to erode the unglaciated lower valleys. The rivers form a broad net collecting the rainwater and melting ice and snow of the headlands—two dozen streams of cold, clear water flowing parallel to one another down from the mountains. The San Joaquin River collects the southern channels and the Sacramento River collects the northern channels. These two great rivers run parallel to the mountains through their broad valleys and into a deep embayment of the Pacific Ocean.

These river canyons were once a paradise. The earliest written descriptions are of water clear as crystal and swarming with salmon and trout. White ash, alder, maple, laurel, and wild grapevines shaded honeysuckle, ferns, and mosses. American dippers and kingfishers worked the pools. Deer were plentiful along the bottomlands. John Muir wrote reverently of the fountain canyons of the Merced River and of the natural gardens and water music of the Tuolomne.[64]

The region supported a diverse array of Native American tribes: Wintu, Sinkyone, Pomo, Maidu, and others lived as hunter-gatherers. They fished for salmon, trout, and eel in the rivers and along the coast. Bear, elk, deer, and smaller mammals supplemented a plant diet that featured acorns. The tribes used controlled burns in the autumn to create better spring forage for the grazing animals they hunted. Estimates are that California had a population of 310,000 people before the Spanish arrived.[65]

From 1769, the Spaniards extended their system of missions and presidios north from the deserts of Mexico and Arizona. They brought with them diseases such as smallpox, measles, and diphtheria, and the population of Native Americans declined by at least 50 percent. The Spaniards converted many of the remaining Native Americans to Christianity, encouraging the natives to settle in agricultural communities around the mission churches and to work for the Spaniards developing cattle ranches in the fertile basins below the Sierras. The Anglo-American explorer Jedediah Smith reached California from the United States only a few decades later, in 1826. Within seven years Anglo settlers followed him. John Sutter, an emigrant from Switzerland, settled in the area in 1839. U.S. Army explorer John C. Frémont reached the area during his 1844–45 expedition, at about the same time as the first large wagon train from the United States. Other wagon parties quickly followed, including the ill-fated Donner party in 1846, and the influx of new settlers led to a series of military skirmishes between U.S. and Mexican forces during 1846 and 1847. The United States won, and its troops occupied the region. The area nonetheless remained part of Mexico until the 1848 Treaty of Guadalupe Hidalgo ceded territory west from Texas and north to Oregon to the United States for $15 million. There were approximately 400 Anglo-American settlers in California in

1848, and the region had a population of about 14,000 thousand Mexicans and 150,000 Native Americans. Then, while building a sawmill along the South Fork of the American River on the site of an old Maidu village, Sutter and his laborers accidentally discovered placer gold on January 24, 1848.[66]

The rush began in May 1848. By 1849 gold was found on several other rivers draining the Sierra Nevada, and gold camps were established on all the rivers by the 1850s. The Anglo population jumped to nearly one hundred thousand in 1849 and three hundred thousand in 1853. The Native American population declined to twenty thousand by the end of the 1850s.[67]

Within three years of the original gold discovery, more than $60 million in gold was taken from surface placers. In five years, eager miners built five thousand miles of ditches and flumes to bring water to their claims and work the sediments. More than one hundred million ounces of gold—a third of all the gold mined in the United States up to the twenty-first century—came from the Sierra Nevada. Between 1851 and 1855 California produced approximately 175,000 pounds of gold a year. This peaked at 200,000 pounds in 1853, which was half of the gold production in the world at the time. The United States continued to produce more than 40 percent of the world's gold output from 1851 to 1860, and most of this came from California. The great majority of the gold mined—89 percent—came from placer deposits until 1875.[68]

The United States rushed to admit California into the Union in 1850, the year that the resourceful miners developed the technique of diverting water via canals. They had already developed the "long tom"—an enlarged rocker eight to fifteen feet long—in 1849. The miners also discovered that mercury could be used to recover gold. Four years later Edward Matteson invented the nozzle and hose that permitted hydraulicking. Using pressurized water to blast down through thick deposits of sediment, hydraulic miners could work large deposits more rapidly and thoroughly. Within months of discovery of any site, the miners working a claim went from individuals to teams so that they could supply the water and manpower necessary for a long tom or hydraulicking. The first mining companies were joint stock ventures owned and administered by miners. Then the miners discovered the gold-bearing gravels of the older river network, now perched four hundred to five hundred feet above the nearby modern valley bottoms. As it became more common

Top: Posing with relatively primitive mining equipment, miners display their work near Lincoln, California, circa 1849. Bottom: Miners pose with a rocker box, wheelbarrows, picks and shovels, and gold pans. One miner shows a nugget in a pan. (Both photographs courtesy of the Bancroft Library, University of California, Berkeley.)

Two views of riverbed mining at Grizzly Flats, El Dorado County, California, circa 1850. (Courtesy of the Bancroft Library, University of California, Berkeley.)

View of a pit along the Middle Fork of the American River in California, Monte Rio Mining Company, 1903. (Courtesy of the Bancroft Library, University of California, Berkeley.)

Todd Canyon flume supplying water to the Monte Rio sites. (Courtesy of the Bancroft Library, University of California, Berkeley.)

Detailed view of pumps set in the pit of the Middle Fork of the American River. (Courtesy of the Bancroft Library, University of California, Berkeley.)

to mine placers in ancient river deposits away from a convenient water supply, companies were formed to build ditches and deliver water. Some of these companies were commercial operations; others remained in the hands of the miners themselves.[69]

A quarter of all the gold mined in the Sierras came from hydraulicking. Sediment began to pour downstream with the river water, and even the mouth of the Golden Gate off San Francisco turned brown. In 1850 there was a flood in Sacramento, a foreshadowing of floods in 1851 and 1853, as well as the 1861 and 1862 floods that breached the town's protective levees. In some narrow canyons, rivers flowed on top of 150 feet of mining sediment ready to be swept down into the valleys. Sediments rushing down the mountain rivers came to rest in the shallower channel reaches at the base of the mountains, burying thousands of acres of rich bottomland being settled by rapid influxes of farmers. The upper branches of mature oak trees poked out of the floodplain sediments of

Top: Piping, or hydraulicking, at the North Bloomfield hydraulic mine, Nevada County, California, circa 1871. (Courtesy of the Bancroft Library, University of California, Berkeley.) Bottom: Contemporary view of the North Bloomfield hydraulic mine site, now Malakoff Diggins State Park.

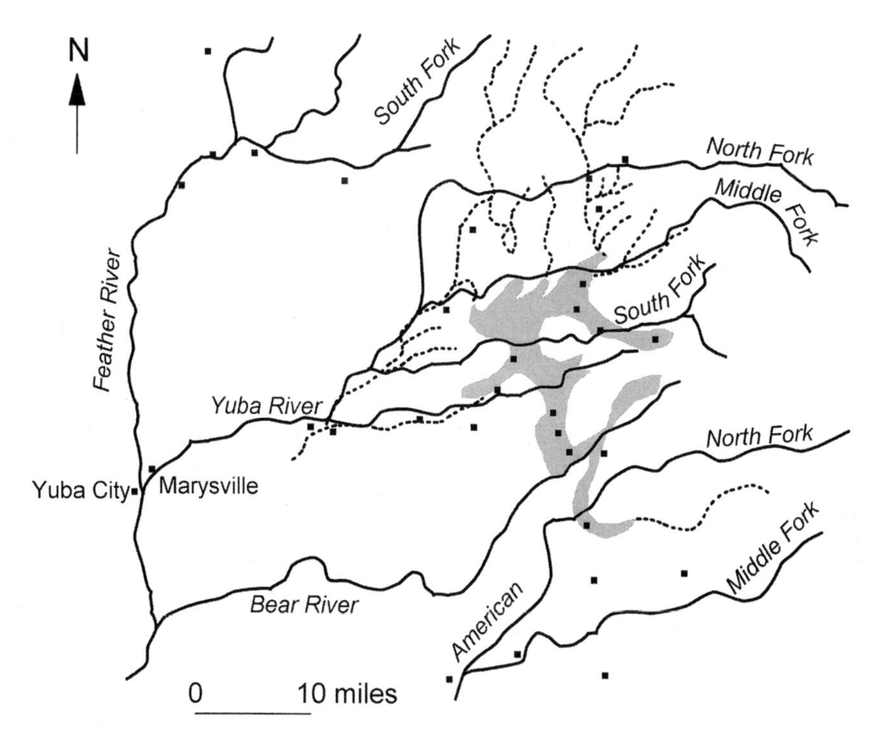

Detail map of mining sites (small squares), modern rivers, and placer deposits along ancient river courses (shaded areas and dotted lines) near the junction of the Feather, Yuba, and Bear Rivers in California. (After Holliday, 1999, *Rush for riches: Gold fever and the making of California,* Oakland Museum of California and University of California Press, Berkeley.)

newly raised rivers. The downstream portions of the bed of the Bear River rose as much as 16 feet. In 1868 the channel beds of the Feather and Yuba Rivers at Marysville were higher than the city streets. The floods grew worse as the channel beds rose. The years 1862, 1875, and 1881 were flood years. Debris tailings buried a large portion of Marysville when the city levees breached during the 1875 flood. Along some rivers even the annual floods drove the farmers out and forced the towns to build protective levees. The floodplains of the Yuba, the Bear, and the North Fork American—the rivers at the epicenter of the hydraulic mining frenzy—were destroyed as sediment up to 100 feet deep choked the river canyons for miles. Sediment accumulated for forty miles along the Feather River downstream from the mining camp of Oroville, burying productive farms. An estimated 100 million cubic yards of sediment

Contemporary upstream (top) and downstream (bottom) views of the Yuba River near Marysville, California. Extensive sediment derived from placer mining remains along the river, and the streambanks host several aggregate mines.

Nineteenth-century view of tail sluices, Yuba River, California. (Courtesy of the Bancroft Library, University of California, Berkeley.)

accumulated in the Feather River, 684 million in the Yuba, 254 million in the Bear, and 257 million in the American River; together, approximately thirty-nine thousand acres of land were buried in sediment. In a year and a half, hydraulic mining along the Yuba alone produced enough sediment to fill the Erie Canal. By 1878, eighteen thousand acres of farmland along the river were covered as tailings spread ten miles into the Great Central Valley. The elevation of the Sacramento River in the Great Central Valley rose 7 feet. In 1880, 1,146 million cubic yards of sediments were added to the bays of San Francisco.[70]

Finally, the downstream settlers had had enough. Legal injunctions began to limit the dumping of coarse mining debris in 1882. The 1883 collapse of a large mining debris dam released 650 million cubic feet of water and caused extensive damage. In 1884, beleaguered farmers and townspeople persuaded the U.S. Circuit Court to prohibit hydraulic mining that discharged sediment into rivers in the Sierra Nevada.[71]

The sediment remained in the rivers draining the mountains, however. Recognition of the problems associated with the excess sediment prompted the federal government to commission studies of the extent and movement of the sediment and of means to mitigate sediment-related hazards in navigable waters. Among the scientists who worked in the region was G. K. Gilbert of the Geological Survey. Gilbert interpreted field and experimental observations to indicate that mining sediment formed a wave moving slowly downstream along the Yuba and other rivers. Working in 1914, Gilbert predicted that most of the sediment would move past Sacramento within about fifty years (circa 1967) and that the channels would then return to premining conditions. Subsequent studies in the 1980s and 1990s indicated that Gilbert's predictions of rapid sediment evacuation were too optimistic in some cases. Extensive volumes of mining sediment remain in the upper Bear River, and more than 90 percent of the mining sediment reaching the lower portions of the river basin remains in storage there beneath channels and floodplains. The Yuba River apparently did flush much of the mining sediment from the streambed and stabilize by 1950, as did the streambed of the Sacramento River by 1930. Most of the mining sediment deposited on the floodplains of these rivers remains there, however. Sediment movement along all of these rivers was influenced by the construction of levees that channelized stream flow and restricted exchanges between the channel and floodplain, and by the construction of dams that stored sediment upstream.[72]

From 1850 to 1950, the river channels draining the western Sierra Nevada remained unstable. The channels continually alternated between filling in response to sediment increases produced by mining or by floods flushing sediment downstream, and downcutting in response to sediment decreases associated with the construction of dams, the end of mining at a site, or sediment storage in floodplains. As late as 1995, a National Academy of Sciences report on flood risk management in the American River basin gave special attention to the effects of mining-related channel instability on flood hazards.[73]

Metal contamination accompanied the sedimentation problems. As mentioned previously, mercury was widely used in processing Sierra Nevada placer deposits. Drainage from lode mines and tailings piles added other toxic materials to the rivers. The dispersal of these toxics through mountain streams is very complex. The mode of transport, whether in solution in the stream flow or attached to particles of fine sediment, depends on the rate of chemical reaction. This rate in turn depends on the water velocity, chemistry, and temperature, as well as the turbulence of flow. The rate of chemical reaction is also influenced by characteristics as diverse as the amount of mixing between surface and groundwater; the intensity of sunlight on the stream water; and the presence and type of sediment, especially organic materials and clay. If the toxic material is transported in association with sediment, it may be partitioned along the river as a function of distance from the source, depositional sites for fine sediments, or frequency of stream flows capable of remobilizing and transporting fine sediment in storage. An example comes from a 1997 flood along Nevada's Carson River that remobilized mercury-contaminated sediment deposited along the river after Comstock mining operations during the mid- to late-1800s. The erosion, redeposition, and storage of sediment and sediment-bound mercury were greater along the segments of the river with shallow gradients and wide valley floors than along steep, narrow portions of the river. However, high concentrations of a toxic material in sediment are not necessarily linked to high toxic concentrations in the nearby waters. Boulder Creek in Arizona supports a robust aquatic community despite high concentrations of copper, lead, and cadmium in the streambed sediments because alkaline stream conditions prevent large quantities of metals from entering the water.[74]

The dispersal of toxic materials through the river ecosystem also depends on biological interactions. Biological responses are difficult to predict because some toxic materials react differently when other toxics are present. Elevated levels of zinc and copper together are more toxic than equally elevated levels of only one. As toxic materials are transported downstream, they cycle among biotic and abiotic components of the stream ecosystem. These multiple pathways of transport and broad dissemination make it extremely difficult to trace and control the toxic materials.[75]

Individual species and types of organisms react differently to inorganic contaminants introduced by mining. Among aquatic insects, some species such as heptageniid mayflies, or some insect functional groups such as scrapers and predators, are highly sensitive to heavy metals. These organisms cannot survive in rivers affected by mining sediment. Other species such as caddis flies, or functional groups such as shredders and collectors, are relatively tolerant of metals. Invertebrate abundance, species richness, and community structure are thus strongly influenced by the presence of heavy metals associated with mining. A study comparing the death rate of fish exposed to four metals associated with placer mining found that copper was most toxic to all species and life stages, followed in descending order by zinc, lead, and arsenic. Sensitivity was greatest in juvenile fish rather than fry or adults, and Alaskan Arctic grayling were more sensitive than coho salmon and rainbow trout.[76]

Other studies have evaluated the impacts of overbank sedimentation of metals during floods. Vegetative diversity and density decrease once the concentrations of trace metals in the floodplain soil cross a threshold. This threshold varies for different vegetative species and different metals. These effects persist for decades after the metals are released into the environment.[77]

In the rivers of the Sierra Nevada, the native fish species are still present in most cases, but their numbers are greatly reduced. Forty kinds of fish are native to the Sierra Nevada, and eleven of these are found only in the range. The natives include species of salmon, trout, chub, and dace. Anadromous fish such as the chinook salmon are no longer present in most of the rivers on the west side of the mountains. Native Americans harvested an estimated eight and one-half million pounds or more of chinook salmon annually from the Sacramento and San Joaquin Rivers. Since the 1850s, the salmon have declined to small fractions of their previous numbers despite substantial human investments in managing these fish. Salmon had their last healthy run on the Sacramento River in 1852. About that time an observer described the turbid waters of the Yuba River as having "once contained trout, but now I imagine a catfish would die in it."[78]

Thirty species of nonnative fish were introduced to, or invaded, most of the rivers of the Sierra Nevada. These fish include shad, carp, catfish, bass, perch, and crappie. Mining, logging, grazing, channelization, and dams and diversions already impacted the native fish; the introduced fish provided further stress. As the mountain river fisheries shifted from native to introduced fishes, the listing began. Six native fish species are listed as threatened or endangered. Twelve species are considered to be of special concern and potential candidates for listing. Four species are in decline in the Sierra Nevada but stable elsewhere. Only eighteen native species have stable or expanding populations.[79]

The amphibians along rivers in the Sierra Nevada have also declined.

Upstream view of a bedrock gorge along the South Yuba River near the Malakoff Diggins historical mining area, California.

Among the introduced fish were nonnative species of trout released in high-elevation rivers and lakes that historically did not have fish. The trout became the top predators and helped to reduce or locally extirpate native ranid frogs. These frogs—yellow-legged frogs, red-legged frogs, leopard frogs, and Cascade frogs—are extremely sensitive to environmental change. When placer mining destabilized the stream channels, streamside vegetation declined and the water in the shallow, unshaded rivers grew too warm for the frogs and their eggs. Narrow, stable reaches of the rivers dominated by bedrock now have streamside vegetation, and frogs. Those portions of the rivers with wide, braided channels still choked with mining sediment are largely unvegetated, and frogs no longer sing from cool, shadowy pools.[80]

Birds That Swim in Rivers

Geologists have facetiously referred to rivers as the gutters down which flow the ruins of continents because the sediments carried by rivers represent the erosional remnants of landmasses. If rivers were only this, then the disruption and contamination associated with placer mining would not matter. But of course rivers are much more than gutters. They are intricate, complex, self-sustaining ecosystems that exert a much greater influence on all aspects of terrestrial and marine systems than the surface area covered by rivers would imply. We care about rivers as ecosystems because we care about other living organisms, and about our own lives. Thoughtful examination of a single riverine species impacted by placer mining gives a specific face to the myriad effects of mining discussed in this chapter.

The American dipper (*Cinclus mexicanus*) is a river bird. It is born to the sound of rushing water in a domed nest on a cliff ledge, behind a waterfall, or on a large boulder in the midst of a stream. The ovenlike, insulated nest reduces heat loss while the female dipper incubates her four or five eggs for just over two weeks. She must take advantage of strategies to keep herself and the eggs warm, for the dipper is a bird of cold rivers. From the chilly coastal streams of Arctic Alaska, the dipper's range follows the mountains southward all the way to western Panama. Across this huge swath of territory the dipper is selective, living only along swift, rocky streams with cold water capable of supporting the aquatic insect larvae that make up most of the bird's diet. To get these larvae the dipper goes into the river, not with a casual dip of the head made by a duck feeding in quiet water, but walking the bed of swiftly flowing mountain streams or diving into pools and swimming underwater. The bird becomes less active during very cold weather, particularly if the stream is largely frozen over, but it is quite capable of plunging its little body off an ice ledge and into frigid water.[81]

Several adaptations allow the dipper to be so at home in its paired worlds of water and air. The bird has scales that seal its nostrils while underwater and a large oil gland at the base of its tail that provides waterproofing for its feathers. The dipper is densely feathered, with a heavy covering of down between its feather tracts, and even feathers on its eyelids. Only its legs and feet are uninsulated, and it is through these that

A dipper on a streamside rock. (Courtesy of Peter LaTourrette.)

the dipper loses body heat when air temperatures rise. The bird cannot survive at air temperatures above 97 degrees Fahrenheit because its body overheats, but it can maintain a normal body temperature when the outside air temperature drops to −22 degrees Fahrenheit. The dipper also has a low metabolic rate and extra oxygen-carrying capacity in its blood. This allows it to remain below the surface of frigid stream waters as it searches for insects. And finally, the dipper is able to see equally well in air and in water. The muscle encircling the dipper's pupil is much more developed than that of otherwise comparable songbirds. This suggests that the curvature of the dipper's lens is changed by the pressure this musculature exerts on the lens, allowing the dipper's eyes to better accommodate changes between air and water.[82]

The dipper is seldom far from water. Once a young fledgling leaves its nest, it may cross a drainage divide to another stream. Beyond that, the bird is unlikely to be out of sight of a river for the rest of its life. Both male and female dippers establish and defend territories, moving upstream during spring and summer, or simply remaining along the same stretch of river throughout the year. A good habitat has up to six dippers per half mile of river.[83]

There is likely to be a large turnover among these six dippers. One study of dippers in the Colorado Front Range found that nearly 90 percent of dipper eggs never reached breeding age, while only about 20 percent of adult dippers survive from one year to the next. Winter is a time of major loss, although there is little information on what kills the dippers. Predation probably accounts for some. The chunky little gray bird may be eaten by brook trout, snakes, mink, martens, skunks, weasels, water shrews, or wood rats, as well as by birds such as sharp-shinned hawks, Stellar's jays, American magpies, or Clark's nutcrackers. Disease accounts for some dipper mortality. Floods destroy nests built too close to the water, and starvation, particularly during winter, likely plays a role.[84]

Although the dipper eats small fish, fish eggs, flying insects, and even plants, most of its diet comes from stone fly and mayfly nymphs and caddis fly larvae. These creatures live on or under rocks in the streambed, and the dipper sometimes moves rocks in search of the insects. It takes a lot of insects to keep an energetic little dipper going. One study in Montana identified fifteen hundred insects in the stomachs of twenty-six dippers. One bird had a belly full of 322 caddis fly larvae.[85]

Upon finding something, the dipper returns to the water surface to eat. Small, unprotected creatures are easy to consume. Caddis fly larvae must be extracted from their pebble cases. Fish or other larger prey are subdued by being shaken or slammed against a rock, sometimes for ten minutes. Then the dipper is off again on its rapid flight, skimming just above the water surface. Alighting briefly on a rock to preen and groom its feathers, the bird bobs its body up and down at a rate of nearly once a second before plunging once more into the flow.[86]

The dipper enchants many human observers. Writing in 1894, John Muir compared the dipper's piercing calls to water music: "[H]is music is that of the streams refined and spiritualized. The deep booming notes of the falls are in it, the trills of the rapids, the gurgling of margin eddies, the low whispering of level reaches, and the sweet tinkle of separate drops oozing from the ends of mosses and falling into tranquil ponds."[87]

The sweet songsters are also a river version of a miner's canary. Dippers are found on channels from tiny creeks to large rivers, if the water is clean. A study of dippers along Montana's Rattlesnake Creek found none on the contaminated portion of the Clark Fork River into which Rattlesnake Creek drained. Unfortunately, the dipper is not a foolproof discriminator. If aquatic insects are present in the cold, clear waters of a stream, dippers try to live there.[88]

Decades of nineteenth-century placer mining in the upper Arkansas River of Colorado produced elevated concentrations of lead, copper, cadmium, and zinc dissolved in the river water or attached to river sediments downstream. The invertebrates inhabiting these portions of the river are metal-tolerant species such as some chironomids and caddis flies with the ability to accumulate toxic metals. This accumulation is passed on to their predators. Trout in the river have reduced biomass and population density and lower survival of adult fish. Dippers in the river show more subtle effects. Lead and cadmium concentrations in the invertebrates correlate with concentrations of lead and cadmium in samples of dipper blood, liver, and whole carcasses. The levels of metallothionein in the birds' livers also correlate with cadmium concentrations in liver and carcass samples taken from both adults and nestlings. Metallothionein is a protein induced by metals. It acts to bind the essential metals copper and zinc, as well as toxic metals such as cadmium and mercury, and it helps to protect an organism against toxic metals. The protein's presence in the dippers is another indicator that cadmium and lead are accumulating in the birds through the food chain.[89]

A further indication of toxic metals accumulating in the dippers is reduced levels of erythrocyt δ-aminolevulinic acid dehydratase (ALAD) activity. ALAD is a rate-limiting enzyme in the biological production of heme, which is a component of hemoglobin, the oxygen-carrying substance in the red blood cells of vertebrates. The dippers' exposure to lead is reflected by depressed ALAD levels. Lead adversely affects all body systems and inhibits enzymes such as ALAD, which all cells require. In dippers, ALAD activity is inhibited by 50 percent at levels of 145 parts lead per billion parts tissue. This inhibition in turn translates into lower hemoglobin levels, impairing the bird's ability to dive for its food in the waters of an icy-cold stream. As the author of the study on the Arkansas River wrote: "Early departures from health in animals are not apparent as overt disease, but are associated with the initiation of biochemical, physiological and/or behavioral compensatory responses. When such responses are activated, the survival potential of the organism may already have begun to decline because the ability of the organisms to mount compensatory responses to new environmental challenges may have been compromised."[90] In other words, the dippers are not dying outright, and they are not obviously diseased, but the resilience of their bodies is insidiously compromised.

The dippers along the Arkansas River inhabit a stream where mining ended more than a century ago. Those living along the clear, cold mountain streams draining the western flank of the Sierra Nevada faced different challenges during the period of active mining. Dippers eat primarily aquatic insects found along cold-water rivers with a substrate of sand, cobbles, and boulders. The birds are thus totally dependent on the productivity of rivers. They occupy linear territories along these rivers and seldom disperse more than six to twelve miles from their birthplace. If the character of a river changes—if its waters grow turbid with sediment and it begins to shift course repeatedly—the dippers living along the river could disperse in search of unoccupied suitable territory along adjacent rivers. But if all the rivers in a region are simultaneously altered, as happened along the western flank of the Sierra Nevada during the 1850s and 1860s, then the dippers may have no refuge. Even small amounts of sediment reduce the density and diversity of benthic invertebrates. Greater amounts of sediment reduce invertebrate densities up to 80 percent, or eliminate invertebrates completely. And if a large segment of a river is disrupted, recolonization of the disturbed reach by invertebrates immigrating from less disturbed areas becomes slower and less likely.

Dippers live today along the rivers of the western Sierra Nevada. Tens of thousands of salmon spawn along the rivers not blocked by dams, and mining-generated gravels appear to provide good spawning sites along some channels. The river ecosystems are resilient, yet they are also poorer for mining. Although we have no scientific census records of animal population numbers or diversity before mining, rivers less impacted by mining can be used as reference sites to suggest what aquatic and riverside communities were like before 1849. However, many of the rivers in the Sierra Nevada are substantially altered by other land uses, such as timber harvest or dams and diversions. A 1998 evaluation of the biotic integrity of watersheds in the Sierra Nevada identified only seven of one hundred watersheds as having high biotic health in terms of abundance and distribution of native fish and frogs. Casual observations suggest that, as with the frogs, dippers are more numerous along recovered reaches of river that are no longer storing or transporting large volumes of mining sediment. It is likely that in general the rivers of the Sierra Nevada have lower densities of aquatic insects than were present historically, and thus lower densities of dippers, although this has not been demonstrated. It is known that the aquatic and riverside communities along the mined rivers are stressed by lingering toxic contaminants such as mercury.[91]

The European American pioneers eagerly and thoughtlessly exploiting America's rivers saw themselves as the first phase of an advancing civilization that would subdue the wilderness and reclaim it for humans. The miners were pioneers in more senses than they realized, for their narrow-sighted, homocentric, utilitarian view of rivers was enthusiastically taken up by those in succeeding generations as they engaged in commercial activities.

Chapter 4

Poisoning America

Commercial Impacts

Land with a population density low enough to be called a frontier disappeared from the continental United States during the 1890s. By that time, the frontier had already been closed for a century in parts of the eastern and midwestern United States. As the U.S. population continued to grow through reproduction and immigration, diverse regions of the country became increasingly densely populated. When combined with rapidly advancing material technology, this increasing population density was reflected in progressively larger scale, more intensive, more organized use of natural resources by commercial and governmental entities. Activities begun by pioneering individuals or communities often increased in both extent and intensity when conducted on a commercial scale. For example, pulpwood log drives were almost universal on small streams in Maine during the 1950s. Stream channels with log drives were routinely bulldozed to change them into smooth troughs, an activity that increased water temperature and siltation of spawning gravels and decreased pools, cover for aquatic organisms, and the number and diversity of aquatic insects. Maine finally passed a law restricting such destruction in the early 1960s.[1]

Commercial activities were dominated by private businesses specializing in timber harvest, dam construction, minerals extraction, or fishing, to name some that most significantly affected rivers. Because agriculturalists increasingly raised crops and livestock for sale rather than for their own consumption, they too can be considered as businesses.

Simultaneous with increasing commercialization of resource use was increasing urbanization in many parts of the country. An undisturbed watershed with native vegetation can be thought of as a sponge that absorbs and slowly releases snowmelt and rainfall. Precipitation falling on an urbanized watershed is better imagined as pouring a glass of water on a kitchen countertop; the water runs off immediately without being absorbed. Urbanization greatly increases the amount of impervious surface area in a watershed as roads, parking lots, and buildings replace wooded or grassy surfaces. The precipitation rapidly shed from these impervious surfaces carries along a load of dissolved contaminants—metals, pesticides, and other toxic compounds—as it rushes into storm sewers that efficiently direct the runoff into adjacent rivers. The rapid runoff creates flood peaks that are larger in magnitude but shorter in duration. These flashier floods, combined with a lack of sediment entering the stream channels because most surfaces are paved, produces energetic flows that erode the streambed and banks. Cities often respond by trying to stabilize (pave) the streambed and banks, and even the floodplain, further exacerbating the problem and shifting the fast, erosive flows downstream. The effects of urbanization on U.S. rivers continue to grow in the early twenty-first century, but they began as early as the eighteenth century in the eastern United States.[2]

Like the period of pioneer impacts, the period of commercial impacts dominates different periods of time across the country. Industrial activities began to affect some rivers substantially in the eastern United States during the first decades of the nineteenth century. Fish migrations along the major East Coast rivers of the United States are blocked by seventy-eight dams, many of which were constructed during the first decades of the 1800s. Henry David Thoreau, in *A Week on the Concord and Merrimack Rivers*, published in 1849, described how dams in manufacturing cities impounded water so far upstream that farmers supposedly distant from the centers of industry watched as their meadows flooded and were destroyed. Industrial activities impacted rivers in parts of the western and midwestern United States by the middle of the nineteenth century and most of the rest of the country by the early decades of the twentieth century. These types of activities continue to substantially affect American rivers today.[3]

The European American perception of unsettled or natural lands as a threat that needed to be dominated by humans, and thus civilized, continued to govern the use of natural resources through the nineteenth century and into the twentieth. This attitude became less pervasive than it had been during the pioneer era, however, as landscapes and ecosystems were increasingly manipulated for European American desires. During the commercial era, Native Americans were extirpated or confined to reservations. Large predators were hunted to near-extinction or extinction. Wildfire suppression became increasingly effective. Structures were built to transfer water for agriculture and alleviate the threat of droughts. A rapidly expanding transportation network of canals, railroads, and roads facilitated distribution of goods and kept individual communities from being so isolated or dependent on local resources.

As natural lands were perceived as less threatening, they also were increasingly subject to what ecologist Garrett Hardin has described as the tragedy of the commons. If land and resources are not owned by an individual and not protected by a government entity, they are perceived as "up for grabs." Consequently, they may be exploited as rapidly and thoroughly as possible by individuals or corporations who are optimizing short-term profits and have no long-term interest in resource conservation or sustainability. Widespread attitudes that hard work and cleverness, supplemented by greed, could lead to the quick accumulation of wealth and social status facilitated this type of exploitation. Anyone doubting this reality had only to look to the widely admired and influential "princes of industry": steel magnate Andrew Carnegie, railroad baron Edward H. Harriman, oilman John D. Rockefeller, or lumber capitalist Frederick Weyerhauser.[4]

Two facets of the American booster mentality also facilitated commercial exploitation of natural resources. The first was the perception that economic health depended on a rapidly expanding economy, which was best achieved by attracting settlers and businesses to a region. Industrial growth in particular was deemed so vital that the worst side effects of industrialization—air fouled by factory smokestacks or water rendered undrinkable by manufacturing effluent—were tolerated as necessary evils of progress. The second aspect of the booster mentality was that any community could become a major economic center if only it grew rapidly enough. Communities lobbied vigorously for transportation routes such as railroads or highways that would attract industry, and often provided incentives to industry in the form of reduced taxes or regulatory oversight.

By 1900, U.S. industry was dominated by manufacturing centers along the East Coast (New York, Boston, Philadelphia) and the Great Lakes (Chicago, Cleveland, Pittsburgh, Milwaukee, Detroit, Buffalo) and their outlying areas (St. Louis, Cincinnati). The remainder of the country had regional centers such as San Francisco, Seattle, or Denver that relied less on manufacturing and more on commercial mining, timber harvest, or agriculture.[5]

Opposing these tendencies for accelerated commercial exploitation of the American landscape were the start of the conservationist movement and the initiation of federal governmental controls on land ownership and resource use. Most people did not think about conserving a specific resource or landscape until the supply of that resource was in danger of vanishing, or until the landscape represented a last fragment of something once widespread. The extinction or near-extinction of species once too numerous to count, such as passenger pigeons or bison, are infamous examples of incredibly rapid resource consumption. On a smaller scale, communities might face exhaustion of a local resource, such as fish or timber or agricultural water supply, before exploitation of that resource began to be regulated for conservation. For example, commercial fishing on the Great Lakes evolved slowly until the mid-nineteenth century. Fishing then escalated in response to population growth, wider markets, improved transportation, and new techniques of harvest and preservation. Fish were free for the taking until 1857, when fishing became subject to a modest license fee and loosely enforced regulations. By the time Great Lakes fishing reached a climax in 1889, many people were already concerned about the future of the industry. Decline in fish catches was dramatic by the mid-1920s; the nearly 147 million pounds of fish caught in 1889 had given way to just over 89 million pounds in 1928. But whenever fishery conservation was proposed, it was opposed by fishers afraid to spoil profits and eliminate jobs. Although fish catches did again rise in the Great Lakes, many of the commercially attractive fish species such as whitefish, herring, or sturgeon never recovered, and the catch shifted to carp, suckers, yellow perch, and other "rough fish."[6]

Commercial exploitation of some natural resources was limited by the existence of large tracts of public land in the western United States that the federal government could reserve for restricted use. Yellowstone became the first national park in the United States in 1872, its designation aided by lobbying from railroad owners who expected to transport tourists to and from the region. The first national forest lands were reserved in 1891 specifically in response to recognition that timber supplies in the United States would be rapidly exhausted if their use was not regulated. Gifford Pinchot was particularly effective during the 1890s and succeeding decades in publicizing and popularizing concepts of sustainable forestry and government regulation of timber harvest. Sometimes allied with this movement, and sometimes opposing it, were the private organizations founded to preserve particular landscapes or life-forms. The Sierra Club was formed in 1892 to preserve the mountain regions of the Pacific coast, followed a few years later by the Audubon Society (1905), founded to protect birds. These groups subsequently grew into national organizations with a much broader focus on environmental protection.[7]

While industrialists were developing commercial organizations to exploit natural resources, and conservationists were developing governmental agencies and private organizations to regulate resource exploitation, American science was coming of age. During the nineteenth and twentieth centuries, scientists developed conceptual frameworks to explain how different components of the physical and biological environment had assumed their existing patterns and how these components interacted and controlled one another.

European scientists first systematized investigation of the physical environment during the late eighteenth century. Early geologists categorized different types of rocks and developed the principles of stratigraphy that explained spatial relations between rock units. During this period, James Hutton and others began to explore the role of rivers in shaping landscapes, and during the 1830s Louis Agassiz began the work that revealed how extensively Pleistocene continental ice sheets had shaped topography in the higher latitudes. The government-sponsored expeditions to the western United States after the Civil War not only inventoried natural resources, but also provided a basis for ongoing studies by John Wesley Powell and G. K. Gilbert of the roles of river erosion and mountain building in shaping landscapes.

The results from these expeditions were incorporated into the "cycle of erosion" formulated by William Morris Davis in the 1880s. Darwin's theory of biological evolution heavily influenced this evolutionary scheme for landforms, which distinguished young mountain landscapes, mature landscapes of rolling hills, and old landscapes of flat plains. This formed a conceptual framework for approaching any landscape as representing a point along this linear progression.[8]

The cycle of erosion dominated landscape studies until after World War II. Beginning in the late 1940s, the research emphasis shifted from the historical development of landscapes to studies focused on the physical and chemical processes shaping individual landscape features, such as rivers. Process studies used techniques from physics and engineering to quantify landform change over relatively small scales of time and space. From these studies came the idea that landforms could reach a state of equilibrium in which the rate of change became negligible. Process studies also described characteristic relations among flow, sediment transport, and river form that allowed investigators to predict how a river would respond to changes in water or sediment supply. By the late twentieth century, these predictions were incorporated into computer simulations of river drainage networks and individual rivers.

The basic exploration and inventory of North American biological systems continued into the mid-twentieth century as investigators described the forms, life cycles, behaviors, and geographic ranges of both terrestrial and aquatic organisms. In 1890, C. H. Merriam published a system of terrestrial life zones that equated elevational zones with biogeographic realms within the North American continent. Recognition of these patterns encouraged study of climatic controls on plant and animal distributions. As the research emphasis shifted to biological interactions among organisms, the concept of successional and climax communities was developed in the 1930s. Ernst Haeckel and others first proposed the ecological approach of studying organisms in the context of interactions with their physical environments and with each other in the mid-nineteenth century. This approach came to the forefront of biological research during the 1960s and 1970s, when the language of mathematics was more likely to be used than the descriptive style of earlier naturalists.[9]

In river ecology, investigations tended to remain descriptive and to focus on individual species until the 1950s. After the mid-twentieth century, there was a gradual shift toward examining the whole stream community and the interactions among different species and the physical environment. This resulted in the river continuum concept, which systematically described downstream trends in energy flow, community composition, and food webs. According to this conceptual model, river headwaters consist of narrow streams shaded by vegetation. Leaf litter, twigs, and other particulate organic matter dropped into the stream provide the nutrients that fuel the river food web. The aquatic insects are primarily shredders and collectors that process this organic matter, making it available to other invertebrates and to fish. Water temperatures are low, as is biological diversity. Moving downstream, the streams grow wider. Algae and rooted plants growing in the stream provide more of the nutrients for the base of the river food web, and grazing insects that feed on these plants become more numerous. Water temperature, habitat diversity, and biological diversity all reach maximum levels. Farther downstream, shading from riverside vegetation exerts much less influence on the wide channel. Most of the insects are collectors feeding on fine organic matter drifting in the stream flow. Water temperature and biological diversity are both lower than at mid-length along the stream.[10]

By the end of the twentieth century, river research increasingly emphasized the importance of physical and biological disturbances, such as floods, in shaping river ecosystems. The flood-pulse model recognizes the critical links between a river and its terrestrial setting. A natural, predictable flood pulse provides seasonal access to floodplain habitat for aquatic organisms such as insects and fish. The warm, nutrient-rich waters of the floodplain provide nursery habitat for young fish because these smaller fish can evade larger predators more easily in the shallow waters of the floodplain. As the floodwaters recede from the floodplain back into the channel, they may also carry a pulse of nutrients into the channel.[11]

Increased attention to nutrient dynamics in streams led to the recognition that nutrients in a stream do not cycle in place but are progressively displaced downstream. Ecologists also came to recognize the difficulty of predicting river ecosystem response to alteration because of many interrelated factors. The importance of predators in shaping an aquatic community can change with water temperature, for example.[12]

Dissemination of a scientific understanding of rivers to society at large was accomplished by several means during the later twentieth century. Schools at all levels, from kindergarten to university, increasingly emphasized science and environmental education. Mass media in the form of natural history or science television shows, magazines, and public speaking by scientists and by environmental organizations raised awareness and helped lead to the establishment of such public activities as Earth Day. Books in the natural history genre emphasized the dangers of environmental deterioration. Some of these books, such as

Rachel Carson's *Silent Spring* (1962) or Paul Ehrlich's *The Population Bomb* (1968), were widely read. Lands managed by the National Park Service, and to a lesser extent by other federal agencies, featured museums with interpretive displays presenting different levels of information about environmental alteration. Both the national parks and their interpretive centers in particular are an environmental example of the idea that a democracy depends on citizen education and participation.

At the same time that scientists were trying to understand basic physical and biological processes in rivers, and disseminate their knowledge to the public, human activities were substantially altering river processes. Relatively unimpacted reference sites in national parks or national forest wilderness areas or experimental forests became increasingly valuable as standards against which to compare other rivers. Activities such as mining, fishing, or timber harvest, which affected rivers during the pioneer era, became more intense and widespread during the commercial era. And activities that were relatively unimportant during the pioneer era began to impact rivers substantially. Pollution of rivers through commercial activities, whether manufacturing or agricultural, became one of the most widespread and persistent human effects on American rivers.

Rivers That Burn

During the pioneer era, people tended to live in small, isolated communities. Individual dwellings had outhouses and trash pits, and most disposed material was biodegradable, relatively chemically stable, or of such small quantities that it had little impact on the quality of the surrounding surface and groundwater. Farms were sufficiently small and isolated that animal manure and soil eroding from the farmlands probably had a negligible effect on water quality. As population density in a region increased, soil erosion from cultivated surfaces and sewage from humans and animals increased and affected nearby surface waters.

As discussed in the previous chapter, fine-grained sediment carried into a stream can overwhelm the transport capacity of the stream water and lead to deposition along the channel. This changes the grain-size distribution of the streambed and its ability to support aquatic insects, fish eggs, and young fish. The deposition of fine sediment also reduces the volume of pools that form important fish habitat during summer

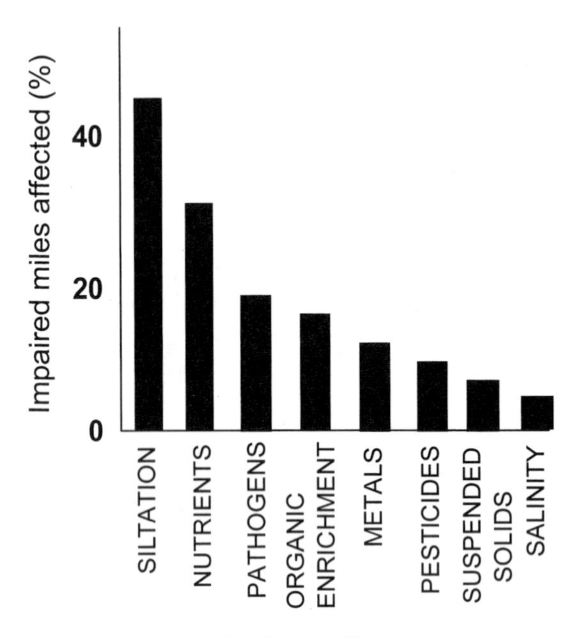

Relative importance of pollutants affecting U.S. streams according to the number of stream miles degraded. (From T. F. Waters, 1995, *Sediment in streams: Sources, biological effects, and control,* American Fisheries Society, Bethesda, Md.)

growth periods and winter cold spells. Even as late as 1990, when numerous pollutants impacted American rivers, the U.S. Environmental Protection Agency still ranked sediment as the most important river pollutant in terms of the number of stream miles degraded in the United States. And sedimentation, along with the upstream river erosion that may be its source, is usually unrecognized by the public.[13]

As also discussed in the previous chapter, excess fermentable organic wastes entering stream waters lead to blooms of microbes and algae that deplete the water's dissolved oxygen and reduce the ability of other aquatic organisms to survive. The history of pollution control along the Illinois River, as reconstructed by C. E. Colten, provides a case study of these effects. The Illinois River drains southwest from near Chicago to its junction with the Mississippi River upstream from St. Louis. By the late 1800s, massive amounts of sewage were flushed from Chicago via the Chicago River into Lake Michigan during floods. Because Chicago took its drinking water from the lake, each flood created a cholera outbreak that led to many deaths. The city decided to divert the Chicago River and the attendant sewage south through a canal to the

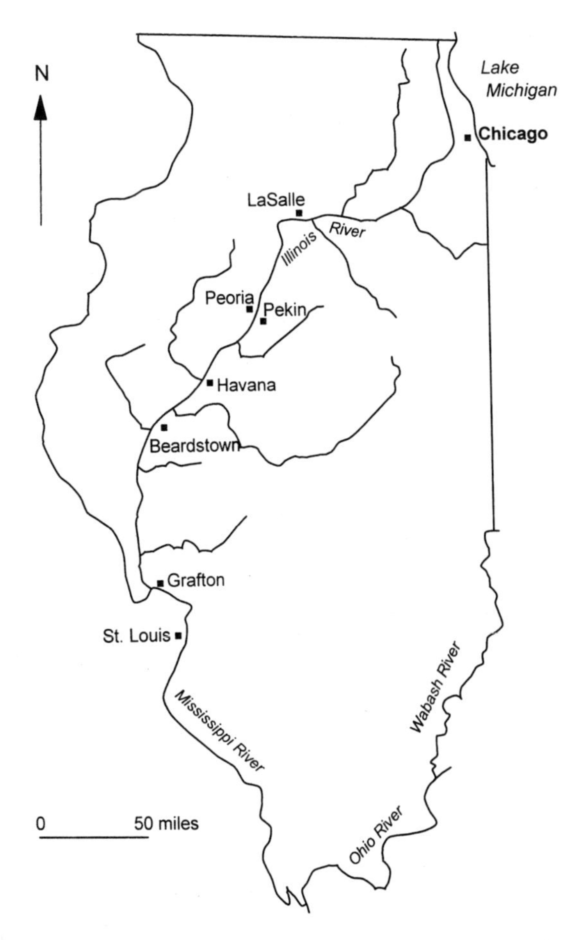

The Illinois River basin. (After C. E. Colten, 1992, Illinois River pollution control, 1900–1970, in L. M. Dilsaver and C. E. Colten, eds., *The American environment: Interpretations of past geographies,* Rowman and Littlefield Publishers, Boston Way, Md., Figure 9-1.)

Illinois River. The annual commercial catch of fish on the Illinois River had increased from six million pounds in 1894 to nearly twelve million pounds in 1900. The Illinois Fish Commission approved Chicago's canal with the belief that the sewage would provide a beneficial food supply to the fish in the Illinois River. The comments of fish biologist Stephen Forbes in the 1910 edition of the *Transactions of the American Fisheries Society* are representative: "We have in Illinois a river . . . which is in many ways one of the most remarkable streams in the country, and in no respect is it more remarkable than in its natural adaptation to the breeding and maintenance of a large and varied population of fishes and other useful aquatic animals . . . we have no reason to suppose that this stream and its adjacent waters have yet reached their limit of economic yield. The effect produced on them by the opening of the drainage canal from Chicago, and the still greater effect due to the introduction of the European carp, are examples of the fact that the original condition of the stream may be largely changed for the better. . . . Evidently this is one of the natural resources of the state and country which should be carefully safeguarded." Forbes's admiration for the "remarkable" Illinois River, and his belief that introduced species and sewage would enhance the river, strike a modern reader as disturbingly contradictory.[14]

The canal opened in January 1900, carrying about 274 million gallons of sewage a day into the river. Downstream cities such as Joliet, LaSalle, and Peoria relied on dilution to keep this enormous influx of sewage from overwhelming their water supply. If dilution failed, they could not enforce their local water ordinances outside of their jurisdiction, or at the source of the problem in Chicago.

By 1908, the fish catch on the Illinois River increased to more than twenty-three million pounds as the sewage fueled more microorganisms and increased flow in the river, and as more fishers worked the river. More than 65 percent of this catch was the pollution-tolerant carp, which made up only 10 percent of the catch in 1894. But after 1908, fish numbers dropped dramatically in the river as a result of both increasing pollution and the loss of wetland habitat along the river corridor.

As early as 1919, scientists proposed that urban sewage was destroying the Illinois River. Increasing water flow down the canal depressed water temperatures and inhibited biological decomposition of the sewage, as well as transporting the sewage farther downstream. By 1915, serious changes in the river's biological community extended as far downstream as Peoria.

By 1922 the urban population of the Illinois River basin reached 3.6 million people, most of whom lived in Chicago. The first sewage treatment plants in the river basin opened in 1922, but more than 140 manufacturers still dumped five hundred million gallons of untreated organic and inorganic wastes into the river every day. This manufacturing input constituted 42 percent of the total waste load by the late 1920s. At that time, only pollution-tolerant organisms populated the upper reaches of the Illinois River, and polluted conditions advanced downstream at the rate of sixteen miles each year. The U.S. Supreme Court ordered the

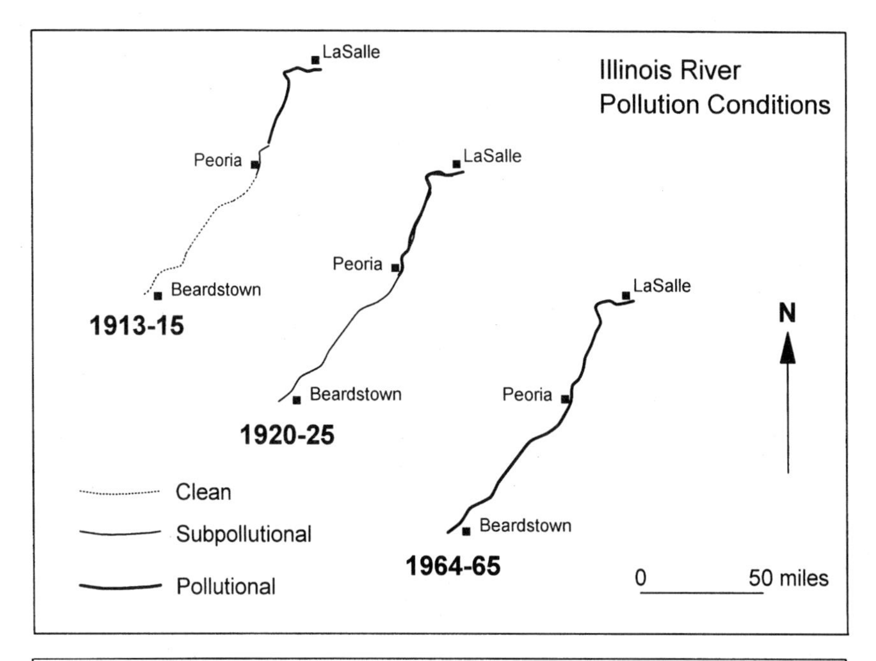

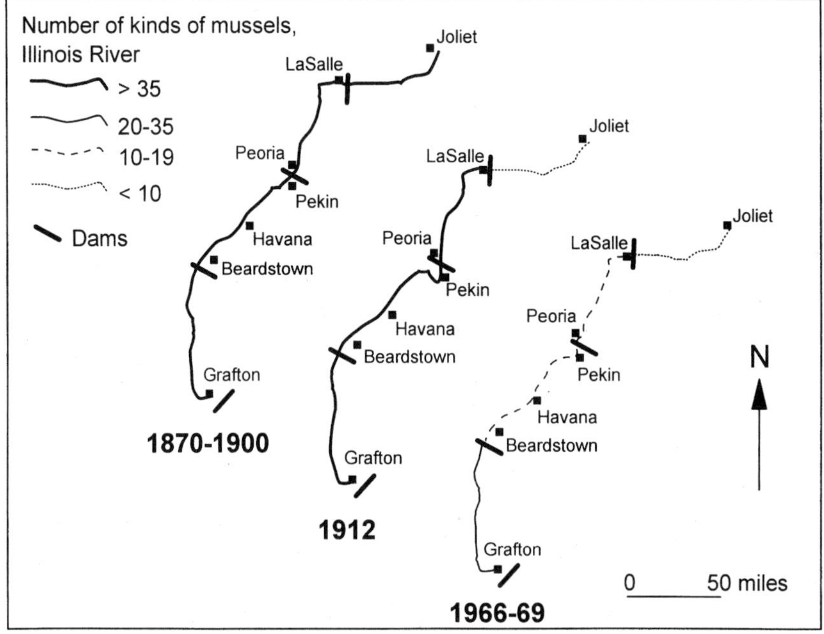

Downstream progression of pollution and extinction of mussels with time along the Illinois River. (After C. E. Colten, 1992, Illinois River pollution control, 1900–1970, in L. M. Dilsaver and C. E. Colten, eds., *The American environment: Interpretations of past geographies,* Rowman and Littlefield Publishers, Boston Way, Md., Figures 9-2 and 9-3.)

construction of further treatment works in 1929, and political agitation for improved water treatment plants continued up to the start of World War II, but the situation changed little.

After the war, the 1948 Federal Water Pollution Control Act provided a framework for inventories and surveys in the Illinois River basin. A 1950 study indicated that more than 110 miles of river were polluted. More than 90 percent of factory effluent received some treatment, but polluters were not prosecuted and problems were increasing. Except for stockyards, industry was not subject to regulation until the late 1950s. By 1957 the proportion of total effluent that remained untreated rose to 34 percent; of 156 sources of factory effluent along the river, 53 received no treatment. The fish catch dropped to 5.6 million pounds in 1950, and to 2 million pounds in 1960. Twenty-five species of mussels disappeared from the river between 1900 and 1966.[15]

The Illinois River remained severely polluted with sewage through the 1960s. Exemptions were allowed for Chicago-area sewage, and water-quality conditions often became extreme before action was taken. The Illinois General Assembly allowed Chicago's sewage to cause the water quality along the river to deteriorate, destroying the commercial fishery and the mussel industry and jeopardizing the water supply of downstream communities, as well as the river ecosystem.[16]

Today, the Illinois River drainage system continues to have problems with organic wastes, organochlorine compounds, other synthetic chemicals, and excess nutrients, but contaminant concentrations are generally below standards for the protection of aquatic life. Chicago detains and treats its urban sewage and storm water, but many other agricultural, municipal, and industrial nonpoint sources throughout the drainage basin continue to introduce contaminants. A dense bottom current of contaminated water still flows north into Lake Michigan along the otherwise diverted Chicago River. Water quality in the Illinois River improves downstream from Joliet, but upstream reaches remain unable to support much diversity of aquatic life despite being heavily treated. The low stream gradient discourages highly turbulent flow and accompanying aeration of the water, so low dissolved oxygen persists. This situation is exacerbated during storm runoff that flushes sediment and nutrients from surrounding areas into the river. Attempts to artificially aerate the stream waters by building vertical drops or installing large screws to create turbulence have had limited success.

Continuing downstream along the Illinois River, sediment becomes the major river contaminant. Of the 13.8 million tons of sediment delivered annually to the Illinois River valley, 8.2 million tons are deposited in the river. This enormous volume of sediment deposition has resulted primarily from changes in the river basin's uplands and river channel. In the uplands, an agricultural shift from wheat to corn and soybeans, and a change in the style of plowing, increased sediment erosion from hillslopes after World War II. Equally important, many of the tributaries of the Illinois River were artificially straightened and deepened to reduce overbank flooding. The rivers historically meandered through enormous wetlands such as the 360-square-mile Grand Marsh, more than 90 percent of which was "reclaimed" for agriculture by straightening the rivers. This process, known as channelization, increases the river's downstream gradient and erosive power and causes massive erosion of the riverbed and banks, as discussed in more detail in the next chapter. Large floods, such as that in 1993, erode huge quantities of sediment from uplands and river channels and flush the sediment to downstream depositional sites.[17]

The sediment being carried into the Illinois River accumulates in backwater zones along the river, the so-called lakes created as water ponds upstream from locks and dams built to enhance navigation between Lake Michigan and the Mississippi River. As sediment accumulates in these sixty lakes, the river must be dredged to maintain the nine-foot depth necessary for barges. This leaves a lot of sediment to be put somewhere. Because much of the sediment is topsoil from farms, experiments are underway to return sediment to farm fields. But contamination of the sediment with metals including lead, nickel, and cadmium from both agricultural and municipal sources makes this problematic. Another approach is to consolidate the sediment into stabilized artificial islands in the river that would also help to narrow and deepen the channel by confining the river flow. The Army Corps of Engineers built the first two artificial islands in 1995, downstream from Peoria, Illinois. One of the islands was still in place as of 2002. A corps spokesperson compared the island construction to the activities of prehistoric Native American mound builders as an example of humans enhancing the subdued topography of Illinois.[18]

Another case study of human impacts to streams from commercial activities comes from the Ohio River. Archeological and histori-

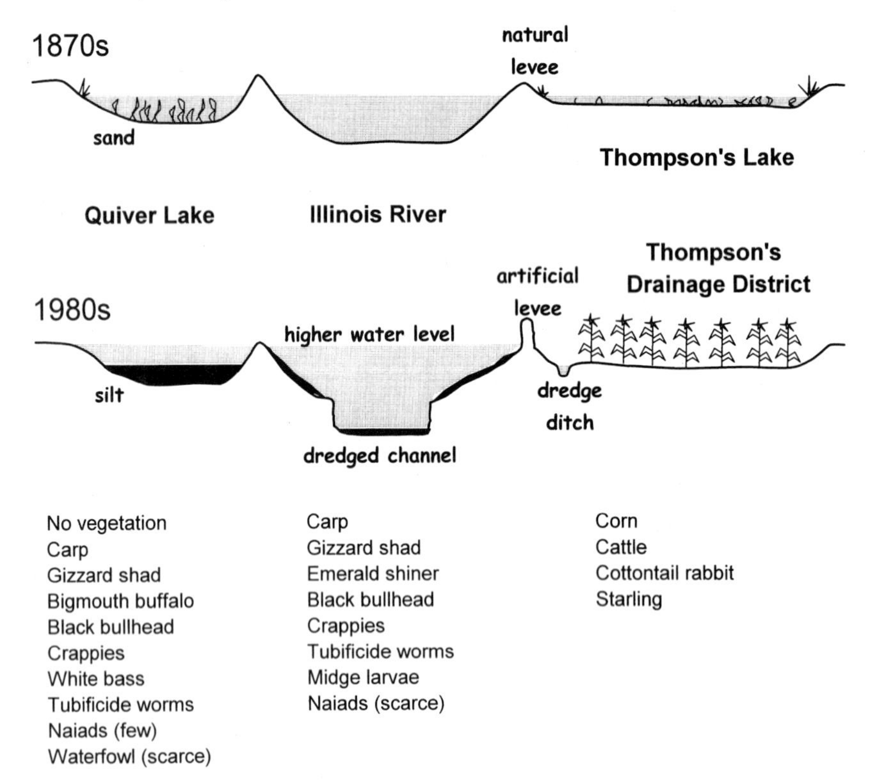

Schematic illustration of changes in the Illinois River between the 1870s and 1980s as a result of changes in land use and channelization of the river. (After R. E. Sparks, 1984, The role of contaminants in the decline of the Illinois River: Implications for the upper Mississippi, in J. G. Wiener, R. V. Anderson, and D. R. McConville, eds., *Contaminants in the Upper Mississippi River,* Proceedings of the 15th Annual Meeting of the Mississippi River Research Consortium, Butterworth Publishers, Stoneham, Mass., Figure 16.)

Barge on the Illinois River, downstream from Peoria, Illinois, July 2002. The river must be regularly dredged to maintain a channel sufficiently deep for barge traffic because of land-use practices in the river's watershed.

cal records from the Ohio River valley indicate abundant fish in the river and much fishing by human residents. Early European American settlers caught pike, buffalo fish, sturgeon, suckers, and freshwater drum. Their journals and letters record thirty- to fifty-pound catfish and twenty-pound perch from the river. To improve navigation and reduce overbank flooding, the Army Corps of Engineers began to remove wood and large rocks from the channel and to construct dikes along the river in 1824. The first dam was built in 1885, and by 1911 there were a dozen more. From 1831 onward, growing industrial development in coal, oil, steel, and meat-packing as well as siltation and sewage from expanding agricultural and urban areas severely degraded water quality in the region. Subsequent estimates put the fish community in the river at sixty species before European American settlement; this community was noticeably poorer by 1900 as several species began to disappear.[19]

Meanwhile, human population in the Ohio River valley continued to increase rapidly, from 750,000 in 1880 to 1.25 million in 1900. Attempts were made to limit pollution in the basin as early as 1908, but these did not really begin to be implemented until 1948. A 1962 study of

Aerial photo mosaic of the Illinois River downstream from Peoria, Illinois, showing narrowing associated with the entry of a tributary stream (middle right) that is depositing large amounts of sediment in the river. Flow along the river is from bottom to top of the photograph.

aquatic-life resources on the Ohio River detected elevated levels of iron and manganese and low pH and low dissolved oxygen, as well as noting that water quality was quite variable with respect to time and location along the river.[20]

As indicated by the examples of the Illinois and Ohio Rivers, increasing population density in a region generally produced some industrialization in the form of commercial processing or manufacturing operations. These in turn produced a wide variety of pollutants released into the air or dumped directly into streams and lakes. Early legislation relating to water quality often focused on urban sewage and largely overlooked industrial wastes. Pollution surveys and investigations published in the early 1970s still focused on sewage and siltation, measuring such parameters as biochemical oxygen demand, dissolved oxygen, temperature, pH, bacteriological concentrations, and insect assemblages but not testing for heavy metals or synthetic chemicals. This was the situation even after the lower reach of the Cuyahoga River in Ohio caught fire in 1969 when a gas-oil spill on the water was accidentally ignited. The spectacle of a river so polluted that it could burn galvanized public attention toward pollution and helped lead to the establishment of Earth Day in 1970.[21]

In 1786 John Heckewelder wrote of the Cuyahoga, "The river itself has a clear and lively current [and] water and springs emptying into the same prove by their cleanliness and current that it must be a healthy country in general." The banks of the river were inhabited by people at least since Mound Builders settled there from 600 B.C. to A.D. 800, and the word "Cuyahoga" came from the Erie tribe's name for the Crooked River. As many as ten thousand Native Americans lived along the lower forty miles of the river before the French reached the area late in the seventeenth century. The British took possession of the region in 1763 after the French and Indian War, and a young George Washington surveyed a portion of the river near the future city of Akron. Moses Cleaveland established a town at the river's mouth in 1796. The town prospered as infrastructure developed, including the Ohio-Erie Canal (1827) between Cleveland and Akron and the first railroad in 1850. By 1900 the region was solidly a part of the manufacturing belt centered on the Great Lakes. Primary industries included the manufacture of steel, automotive products such as rubber and tires, machine tools, petroleum products, chemicals, rubber goods, and wearing apparel. Two and one-half million people lived in the Cuyahoga River basin by 1970, more than 90 percent of them in urban centers such as Cleveland (751,000) and Akron (275,000).[22]

The 1969 fire on the Cuyahoga River is well-remembered today by those who lived through it. But even people living in Cleveland have forgotten that the river caught fire in both 1936 and 1952. The toxic stew that the lower Cuyahoga River became was readily apparent well before 1969. Growing up in Cleveland during the 1930s and 1940s, my father had various summer jobs. One summer he worked in the greenhouses of West Technical High School. The employees were warned to be very careful with the lead-based liquid pesticides they used in the

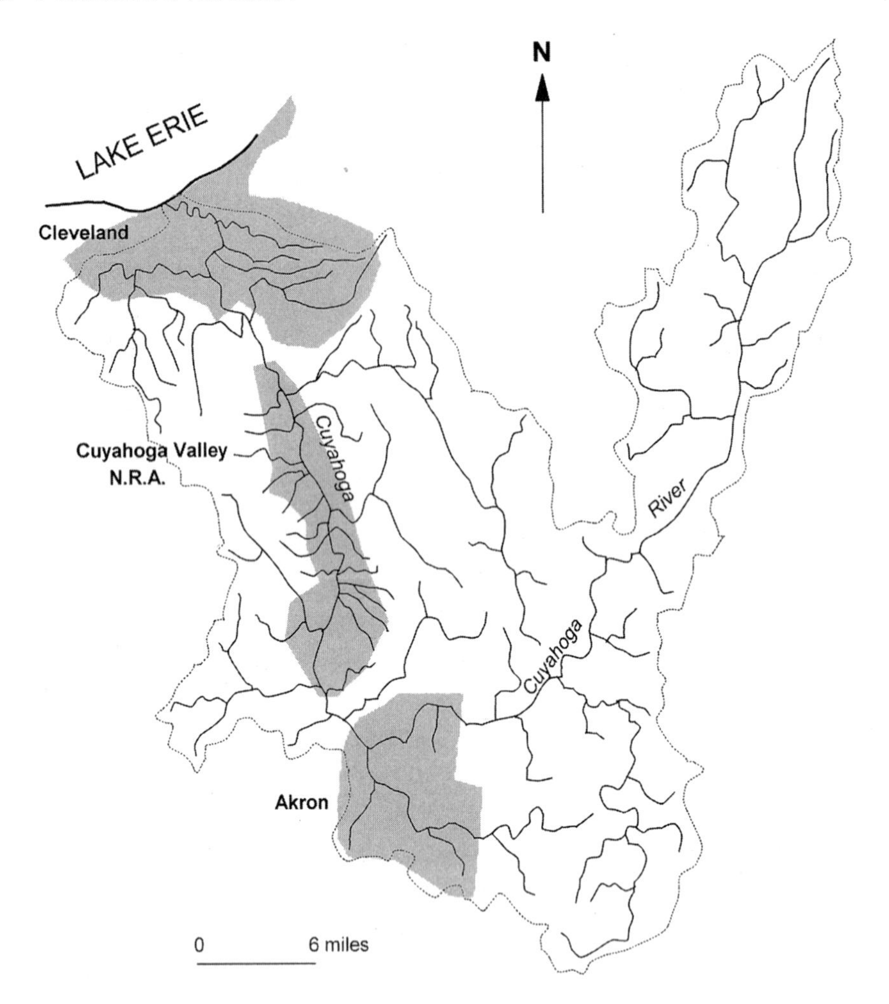

The Cuyahoga River drainage basin in northern Ohio, showing the approximate extent of the cities of Cleveland and Akron and the Cuyahoga Valley National Recreation Area. (After Department of the Army, 1971, *Cuyahoga River basin, Ohio restoration study*, First Interim Report, Buffalo District, Corps of Engineers, Buffalo, N.Y., Plate 3.)

greenhouse, and emergency showers were close at hand. But when one of the workers spilled a container of the pesticides down his leg, he decided to go home to change his clothes rather than using the emergency shower. The young man died later that day. The runoff from that greenhouse went through the storm-sewer system and eventually into the Cuyahoga River or Lake Erie.

Another summer, my father mapped houses for tax revision. Working out of Cleveland City Hall, he became aware of community agita-

Confluence of the east and west branches of the Cuyahoga River about five miles downstream from the river's head. Date unknown. (Courtesy of the Cleveland Public Library.)

tion to have the river cleaned up. Those were the days when a commute across the industrialized portion of the river valley included a stench of chemicals so strong that you were tempted to hold your breath. Metal bridges built across the Cuyahoga in the industrial portion of the river valley corroded to the point of collapse within a decade. Not a sign of life could be found in the river for many miles upstream. Where the Cuyahoga emptied into Lake Erie, there were two pumping plants. One to the east of the river's mouth provided drinking water for the eastern metropolitan area; another to the west of the river supplied the western metropolitan district. During storm runoff, a plume of sewage could be seen extending well into the lake. Yet when citizens protested the river's fouled state, local politicians claimed that it was too expensive to clean up. Cleveland at that time was a very wealthy city.

The Cuyahoga River provides a metaphor for the entire United States. We are a wealthy country, yet we continue to experience horrible pollution because our elected representatives claim that it is too expensive or invasive to regulate the toxins spewed into our environment.

The 1969 ignition of the Cuyahoga River sparked a series of annual public cleanups during which citizens pulled tires and old shoes from the thick muck of the riverbed. Much of the public and governmental response focused on cosmetic changes to the river, although legisla-

THE CLEVELAND WASTE PAPER CO.

Views of the lower Cuyahoga River in Cleveland, circa 1850 (opposite, top), November 1926 (opposite, bottom), and May 1930 (above), showing the changes in extent and type of building along the valley bottom. (Courtesy of the Cleveland Public Library.)

tors renewed pledges to enforce water-quality standards and to acquire land and develop the river's recreational potential. Water-quality analyses published two years after the river fire still emphasized biological oxygen demand, but they also tested for, and found, nitrates, phosphates, sulfates, chlorides, and a wide variety of metals (lead, cadmium, nickel, zinc, chromium, manganese, copper, iron, magnesium, tin, and molybdenum). This was one of the early signs of recognition that pollutants much more insidious than sewage might be accumulating in the nation's rivers.[23]

Subsequent studies chronicle trends across the country in these pollutants. A 1989 study of historical pollution trends between 1880 and 1980 in the Hudson-Raritan drainage basin of New York and adjacent states clearly revealed changing land-use patterns and associated patterns of water pollution. Human population in the basin increased from 3.8 million in 1880 to 17.6 million by 1970. Twenty-four percent of these people lived in rural areas in 1880. The proportion of rural population decreased to 8 percent by 1920 but then rose again to 15 percent in 1980. Emissions and river loadings of metals increased steadily from 1880 to

Fishers along the lower Cuyahoga River in Cleveland, December 1931. (Courtesy of the Cleveland Public Library.)

around 1945 and then fluctuated until they peaked in the late 1960s. These metals came from diverse sources. Iron and steel mills, copper and lead smelters, and primary refineries contributed metals. Fossil-fuel combustion associated with coal-burning and oil-burning electric utility and industrial boilers as well as residential and commercial furnaces contributed. Automotive sources such as exhaust, tire wear, and oil leaked metals into the surrounding waters. Large industrial facilities provided point sources of metals. End-use consumption of batteries, plating, metals, paints and pigments, and electrical and electronic equipment also contributed metals.[24]

The use of the organochlorine pesticides DDT and chlordane in the Hudson-Raritan basin began in the mid-1940s and ended by 1980. Although the 1989 study projected that chlordane runoff would decrease to insignificance during the 1990s, it estimated that DDT runoff would continue well above detection limits into the twenty-first century as soils and airborne transport into the basin continued to release these sub-

Photographs of fires on the lower Cuyahoga River in Cleveland; top: February 20, 1936, printed in the Cleveland News; bottom: in 1952. (Both courtesy of the Cleveland Public Library.)

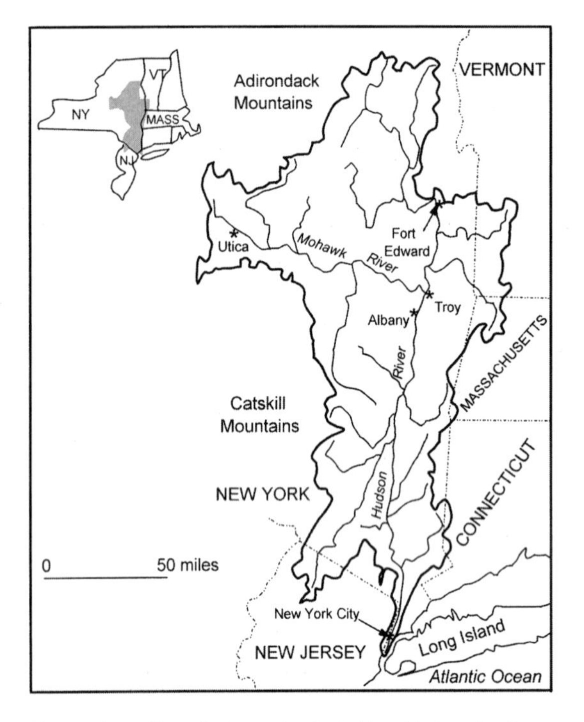

The Hudson River drainage basin in New York and adjacent states. (After P. J. Phillips and D. W. Hanchar, 1996, Water-quality assessment of the Hudson River basin in New York and adjacent states: Analysis of available nutrient, pesticide, volatile organic compound, and suspended-sediment data, 1970–90, *U.S. Geological Survey Water-Resources Investigations Report* 96-4065, Figure 1A.)

stances. These sources of water-quality contamination remained largely unrecognized and unstudied until the late 1970s.[25]

In 1978, nine years after the Cuyahoga River fire, the Connecticut Academy of Science and Engineering published a report that described a whole new class of river pollution in the Housatonic River, which borders the Hudson River basin. The report was titled "PCB and the Housatonic River: A Review and Recommendations." Polychlorinated biphenyls—PCBs—is the generic name for a group of more than two hundred synthetic organic compounds consisting of two linked benzene rings with up to five chlorine atoms attached to each of them. These compounds are extremely stable and slow to chemically degrade under environmental conditions, so they accumulate in the environment. They are destroyed only by rapid burning in an industrial incinerator at tempera-

Table 4.1. Metal Dynamics in the Hudson-Raritan Basin, 1880 to 1980

	Total Emissions (tons)			Total River Loadings (tons)		
	1880	Peak (yr)	1980	1880	Peak (yr)	1980
Cadmium	18	220 (1969)	140	7	120 (1969)	70
Cooper	450	3,300 (1945)	1,800	260	1,700 (1945)	1,200
Lead	1,200	14,000 (1969)	6,900	900	6,800 (1969)	3,000
Mercury	18	330 (1966)	100	14	100 (1966)	70

Source: S. R. Rod, 1989, *Estimation of historical pollution trends using mass balance principles: Selected metals and pesticides in the Hudson-Raritan basin, 1880 to 1980*, unpublished PhD dissertation, Carnegie Mellon University, Pittsburgh, Pa.

tures of 2,000 to 2,700 degrees Fahrenheit, to produce water, carbon dioxide, and hydrochloric acid.[26]

Monsanto introduced PCBs to the United States in 1929. Used extensively in the electricity-generating industry as insulating or cooling agents in transformers and capacitors, they also were used in hydraulic brake fluids, ironing board covers, adhesives and plasticizers, carbonless carbon paper, printer's ink, photocopy toner, paints, sealants, caulking compounds, soaps, and electrical component parts. *Of the 1.4 billion pounds of PCBs manufactured in the United States between 1929 and 1978, 95 percent is expected to enter aquatic environments through wastewater effluents, atmospheric fallout from incinerators, and leachate from landfills.*[27]

In 1968, a supply of rice oil for cooking was accidentally contaminated with PCBs in Yusho, Japan. Nearly thirteen hundred people were hospitalized, twenty-nine died, and many of the survivors suffered permanent effects in the form of skin cysts, respiratory distress, nervous degeneration, abnormal skin pigmentation, and other ailments. PCBs were subsequently linked to death, birth defects, reproductive failure, liver damage, tumors, and a wasting syndrome. They bioaccumulate within an organism, and they biomagnify as organisms pass on their accumulated doses through the food web.[28]

The United States restricted the use of PCBs in 1971, and the Food and Drug Administration used the Yusho incident to set regulatory levels for PCBs in 1973. These levels were subsequently made more strict as scientists realized that PCBs accumulated and that their effects could

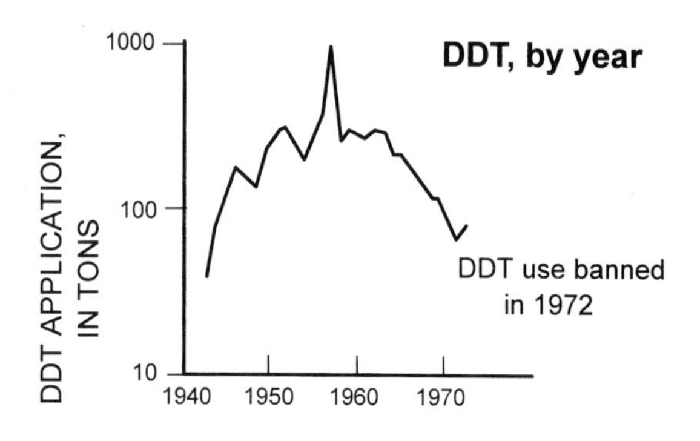

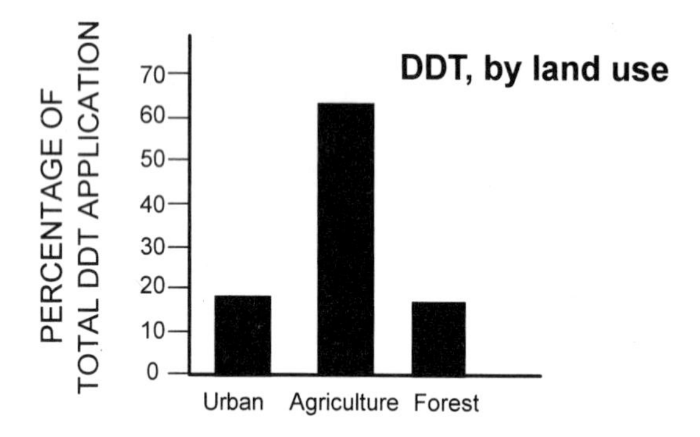

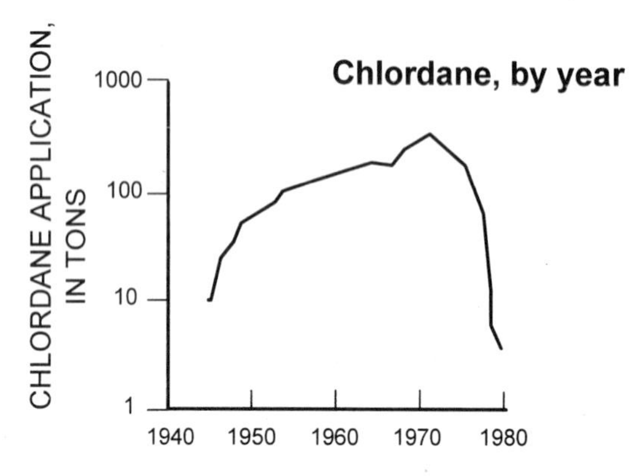

Estimated pesticide application in the Hudson River drainage basin of New York and the Raritan River basin of New Jersey, 1940–80. (After P. J. Phillips and D. W. Hanchar, 1996, Water-quality assessment of the Hudson River basin in New York and adjacent states: Analysis of available nutrient, pesticide, volatile organic compound, and suspended-sediment data, 1970–90, *U.S. Geological Survey Water-Resources Investigations Report 96-4065*, Figure 6.)

be cumulative. The use and manufacture of PCBs was banned in the United States in 1979.[29]

PCBs have an affinity for adsorption onto organic particles in water, as well as being contained in water vapor and adsorbed onto dust particles. Through these pathways, the compounds spread around the world. By the late 1970s they were found in Aleutian birds, Icelandic fish, and the eggs of Antarctic birds, all far from the centers of electrical manufacturing. PCBs accumulated in the Housatonic River over a period of forty years during their release from a General Electric plant in Pittsfield, Massachusetts. The Geological Survey began sampling the river in 1974 and found fish with PCB concentrations greater than five parts per million, the limit set by the Food and Drug Administration for human consumption. In 1977 the river was closed to fishing. People often ignore fishing bans, however. A 2001 survey of private fish consumption revealed that fishers routinely ignored signs warning against or prohibiting fishing in contaminated waters, particularly if the fishers were members of low-income groups or if the sign was written only in English and the fishers did not speak English.[30]

In 1974, PCBs were also detected in 99 percent of the human breast milk sampled in the United States. Twenty-five percent of the milk had concentrations greater than the legal limit of 2.5 parts PCBs per million parts milk and would have been pulled from shelves if it had been a commercial formula. The manner in which our society creates and disseminates synthetic chemicals has transformed one of the most intimate human connections into inadvertent contamination.[31]

The growing recognition of metal and chemical contamination of American rivers occurred within a changing regulatory climate. Under the Interstate Commerce clause of the Constitution, the federal government has exclusive jurisdiction over the navigable waters of the fifty states; however, Congress has largely considered water pollution abatement and control to be the responsibility of individual states. Water pollution control acts in 1948 and 1956 gave states the primary responsibility for such control, with the federal government providing technical and financial assistance. Each state has its own laws and procedures. Generally, the state prescribes water-quality standards and classifies streams according to these standards. The state requires that this classification be observed as a condition of an entity receiving a permit to discharge wastes. If this requirement is not met, the state can bring

A sign in four languages warns fishers not to consume fish caught along the Housatonic River in Connecticut, which flows through a beautiful valley and has no obvious outward signs of being heavily polluted. (Courtesy of Jean Thomson Black.)

charges for violations under the state's criminal laws. The state also reviews and approves plans and specifications for sewage treatment facilities, inspects these facilities, and periodically samples and tests water quality at a point of discharge.[32]

The Clean Water Act of 1972 eclipsed all former legislation and set a "fishable and swimmable goal" for all waterways. Over the next twelve years the federal government contributed a third of the $310 billion spent to clean up surface waters. Spending on water pollution remained high in subsequent decades. In 1987, for example, public and private entities spent an estimated $37.5 billion on water pollution, and in 1992 the Environmental Protection Agency spent $2.9 billion, largely in the form of grants to states for sewage treatment plants. By 1994, public treatment reduced sewage in American rivers by 90 percent relative to 1970.[33]

The reduction in point sources of sewage represented treatment of what the EPA in 1990 referred to as the most "blatant and easily controlled sources of pollution." As a result, fecal bacteria and biological oxygen demand decreased in many rivers. Other contaminants from nonpoint sources continued to increase, however. Salts were flushed into rivers from winter use of road salt. Nitrates from fertilizers were carried on the runoff from farm fields, suburban lawns, and golf courses. Heavy metals entered stream waters from the burning of fossil fuels. Pesticides from urban and agricultural applications dissolved in surface and groundwaters. Heavy metals and petroleum compounds leaked into groundwater from underground storage tanks such as those likely to be present beneath a corner gas station. Salts carried into rivers with irrigation return flow became concentrated downstream as a result of evaporation and reduced flow levels. Reduced flows also increased water temperatures, reduced dissolved oxygen, and concentrated other pollutants. The United States did not, and has not, come close to meeting the 1985 target date of the Clean Water Act for fishable and swimmable waters everywhere.[34]

The State of Our Nation's Rivers

The U.S. Geological Survey's National Water Quality Assessment (NAWQA) program provided a comprehensive index of national water conditions in the late twentieth century. Begun in 1991, phase one of the NAWQA program (1991–2001) had fifty-nine study units spread across all fifty states, from the entire island of Oahu to the Cook Inlet basin in southern Alaska, the New England coastal basins, and southern Florida. The NAWQA program was designed to enhance understanding of natural and human factors that affect water quality. Within each study unit, teams of investigators designated numerous sites at which they repeatedly collected various samples. Sampling focused on seven categories. Investigators analyzed water chemistry for shallow groundwater beneath urban and agricultural lands, deep groundwater in major aquifers, and surface water in streams. They measured grain-size distribution, extent, and chemical contaminants in streambed sediments. They tested tissue from clams or fish for chemical contaminants. They characterized the abundance and diversity of aquatic insect assemblages. They mapped stream habitat. Sample testing included analyses for nine trace elements such as arsenic, lead, and zinc, which are normally present in minute amounts; thirty-three organochlorine compounds, including PCBs, and 106 pesticides; five nutrients such as nitrogen and phosphorus; and sixty volatile organic compounds such as benzene and ethers. Summaries and national syntheses of the first round of NAWQA analyses began to be published in 1996. They make for alarming reading.

The "Dirty Nine": Trace-Element Concentrations in Streambed Sediments

The trace elements known as the "dirty nine" are arsenic, cadmium, chromium, copper, lead, mercury, nickel, selenium, and zinc. Each is very toxic and accessible to living organisms. All nine are on the EPA's 1994 list of 126 Priority Pollutants. They enter stream ecosystems through both atmospheric deposition and point and nonpoint source releases to surface water. Trace elements enter the atmosphere from natural processes such as volcanic emissions and from human activities, including combustion of municipal solid waste and of fossil fuels in coal- and oil-fired power plants, releases from metal smelters, emissions from automobiles, and burning of biomass. Point sources of trace elements include sludge from municipal sewage plants, effluent to surface waters from coal-fired power plants, releases from industrial uses, and drainage from acid mines. Nonpoint sources include natural rock weathering, agricultural activities leading to runoff of manure and arti-

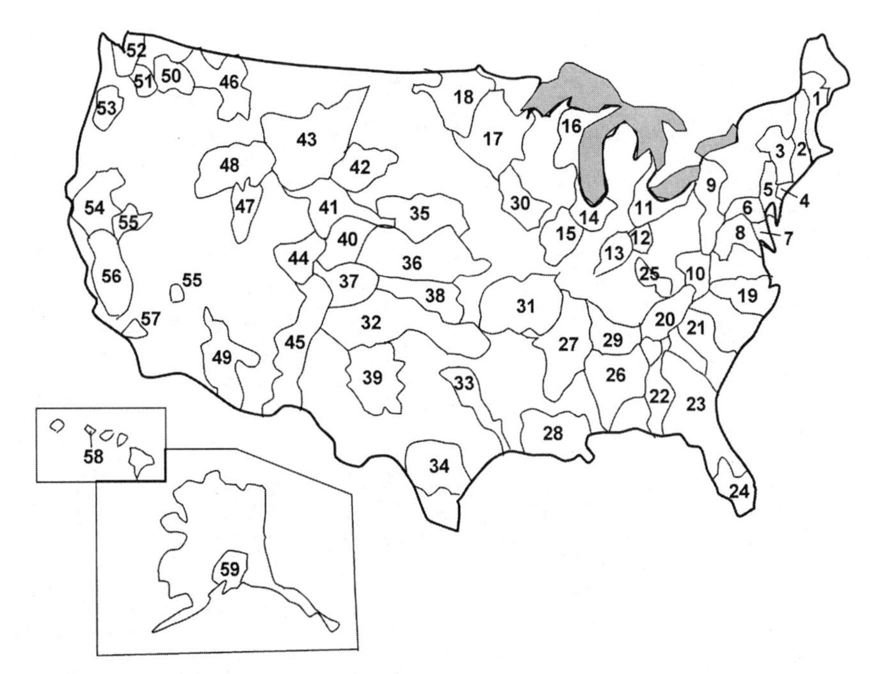

NAWQA Study Units

1. New England Coastal Basins
2. Connecticut, Housatonic, and Thames River Basins
3. Hudson River Basin
4. Long Island-New Jersey Coastal Drainages
5. Delaware River Basin
6. Lower Susquehanna River Basin
7. Delmarva Peninsula
8. Potomac River Basin
9. Allegheny and Monongahela Basins
10. Kanawha-New River Basin
11. Lake Erie-Lake Saint Clair Drainage
12. Great and Little Miami River Basins
13. White River Basin
14. Upper Illinois River Basin
15. Lower Illinois River Basin
16. Western Lake Michigan Drainage
17. Upper Mississippi River Basin
18. Red River of the North Basin
19. Albemarle-Pamlico Drainage
20. Upper Tennessee River Basin
21. Santee Basin and Coastal Drainages
22. Apalachicola-Chattahoochee-Flint River Basin
23. Georgia-Florida Coastal Plain
24. Southern Florida
25. Kentucky River Basin
26. Mobile River and Tributaries
27. Mississippi Embayment
28. Acadian-Pontchartrain
29. Lower Tennessee River Basin
30. Eastern Iowa Basins
31. Ozark Plateaus
32. Canadian-Cimarron River Basins
33. Trinity River Basin
34. South Central Texas
35. Central Nebraska Basins
36. Kansas River Basin
37. Upper Arkansas River Basin
38. Middle Arkansas River Basin
39. Southern High Plains
40. South Platte River Basin
41. North Platte River Basin
42. Cheyenne and Belle Fourche Basins
43. Yellowstone Basin
44. Upper Colorado River Basin
45. Rio Grande Valley
46. Northern Rockies Intermontane Basins
47. Great Salt Lake Basins
48. Upper Snake River Basin
49. Central Arizona Basins
50. Central Columbia Plateau
51. Yakima River Basin
52. Puget Sound Basin
53. Willamette Basin
54. Sacramento Basin
55. Nevada Basin and Range
56. San Joaquin-Tulare Basins
57. Santa Ana Basin
58. Oahu
59. Cook Inlet Basin

Study units of the U.S. Geological Survey's National Water Quality Assessment (NAWQA) program.

ficial fertilizers, releases from wear of automobile parts, and irrigation return flow. New York's Hudson River basin has some of the nation's highest levels of trace elements. In many urban areas of the basin, chromium, lead, zinc, and mercury concentrations in streambed sediments exceeded the levels at which severe adverse effects occur.[35]

The Hudson River basin is not unique in its concentrations of toxic trace elements, however. The NAWQA program found trace elements in all urban areas sampled: "The sum of concentrations of trace elements characteristic of urban settings—copper, mercury, lead, and zinc—was well correlated with population density, nationwide. Median [average] concentrations of seven trace elements (all nine examined except arsenic and selenium) were enriched in samples collected from urban settings relative to agricultural or forested settings. Forty-nine percent of the sites sampled in urban settings had concentrations of one or more trace elements that exceeded levels at which adverse biological effects could occur in aquatic biota."[36]

What are the characteristics and effects of the dirty nine? Cadmium is a relatively rare heavy metal that is concentrated and then introduced to river ecosystems from several sources. Smelter fumes and dusts associated with electroplating, pigment production, plastic stabilizers, and batteries create airborne cadmium that settles on water and soil. Plants take up the cadmium from the soil through their roots and concentrate the metal in their tissues, which humans and other animals can then ingest, passing the cadmium into their bodies. Cadmium also comes from products of incineration of cadmium-bearing materials and fossil fuels, and some cadmium is present in fertilizers. All of these sources introduce cadmium to municipal wastewater and sludge. However, between 50 and 75 percent of the cadmium in the waste cycle comes from batteries.[37]

Because cadmium can rapidly attach to particles of silt, clay, sand, and organic material, river muds may concentrate cadmium by five thousand to five hundred thousand times relative to stream waters. This is not an effective filtering mechanism, however, for cadmium can be detached equally rapidly and become available to living organisms. Organisms accumulate measurable amounts of cadmium from water containing cadmium concentrations not considered hazardous (two-hundredths to ten parts cadmium per billion parts water). Once in an organism, cadmium is a teratogen, which causes developmental changes and abnormalities; a mutagen, which causes chromosomal changes; and a carcinogen, which causes cancerous growths. It concentrates in the organs of vertebrates, especially the liver and kidneys. Although humans gradually excrete cadmium through urine and feces, it may remain in the body for half a century. Ronald Eisler of the Fish and Wildlife Service conservatively estimates that adverse effects on fish or wildlife are either pronounced or probable when cadmium concentrations exceed three parts per billion in freshwater or one hundred parts per billion in the diet. Three parts per billion. That is an incredibly minuscule amount of anything; a little cadmium goes a long way.[38]

Chromium enters river ecosystems from electroplating, metal finishing, municipal wastewater and sludge, tanneries, oil drilling, and cooling towers. Although sensitivity to chromium varies widely among organisms, at high environmental concentrations it is a mutagen, teratogen, and carcinogen. Chromium is not biomagnified through the aquatic food web, and the highest concentrations usually occur low in the food web. Adverse effects occur in sensitive species at as low as ten parts per billion chromium in freshwater.[39]

Unlike some of the other trace elements, which are useful to living organisms at very low concentrations, all measured effects of lead on living organisms are adverse. Lead disrupts survival, growth, reproduction, development, behavior, learning, and metabolism. It enters river systems from storage batteries, pigments and chemicals, cable covering, pipe and sheeting, smelter emissions, and gasoline antiknock additives. In surface waters, lead can be present in solution or as a particulate. It can be mobilized and released from sediments when the water or sediment chemistry change. Lead is toxic to all categories of aquatic organisms, but its toxicity depends on species and physiological state, as well as on chemical and physical variables. Adverse effects on living organisms at lead concentrations as low as one to five parts per billion include reduced survival, impaired reproduction, decreased growth, and high bioconcentration from the water.[40]

Zinc is an essential element to living organisms in small amounts, but human activities have grossly exceeded these amounts in many river systems. Of the estimated world production of 6.4 million tons of zinc, the United States produces about 4 percent and consumes 14 percent. Zinc enters surface waters from electroplaters, smelting and ore processors, mine drainage, domestic and industrial sewage, combustion of solid wastes and fossil fuels, road surface runoff, corrosion of zinc alloys and galvanized surfaces, and erosion of agricultural soils. As of 1980, U.S. rivers were estimated to input 5,400 tons of zinc to surrounding coastal marine ecosystems each year. Zinc interacts with many chemicals, becoming more toxic in combination with lead, copper, or nickel, for example. Zinc mostly enters river sediments, from which individual organisms can accumulate it within their bodies at rates that vary highly among species and individuals. Zinc's primary effect is on zinc-dependent enzymes that regulate RNA and DNA. Zinc accumulates in the bones and pancreas of birds and mammals and in the gills of fish. At high concentrations zinc is teratogenic, and lethal.[41]

Nickel and selenium are similarly damaging to a variety of living organisms. The previous chapter summarized the biological effects of arsenic, copper, and mercury. Probably the most important point to understand with respect to the dirty nine and other trace elements is that river ecosystems serve as collectors and concentrators. What begins as a trace—a barely measurable contribution from any particular source—accumulates in water and sediment as numerous sources contribute. Organisms varying from plants and insects to fish and humans then ingest, store, and further concentrate these elements until they reach toxic levels at which the organisms suffer impaired health or die. Because a river is connected to the entire landscape, any activity within that landscape impacts the river.

Organochlorine Compounds in Streambed Sediments and Aquatic Organisms

We are prodigal in trying to exterminate weeds, insects, and other "pest" organisms, using approximately one billion pounds of pesticides each year in the United States for these purposes. This number has remained relatively constant after growing steadily through the mid-1970s because of the increased use of herbicides, the 1962 publication of *Silent Spring* notwithstanding. Agriculture accounts for 70 to 80 percent of total pesticide use, and 60 percent of agricultural use is herbicides. Insecticides tend to be applied more selectively and at lower rates. Home gardeners use less total pesticides but apply more pesticides and fertilizers per unit area than do commercial farmers. Environmental concerns

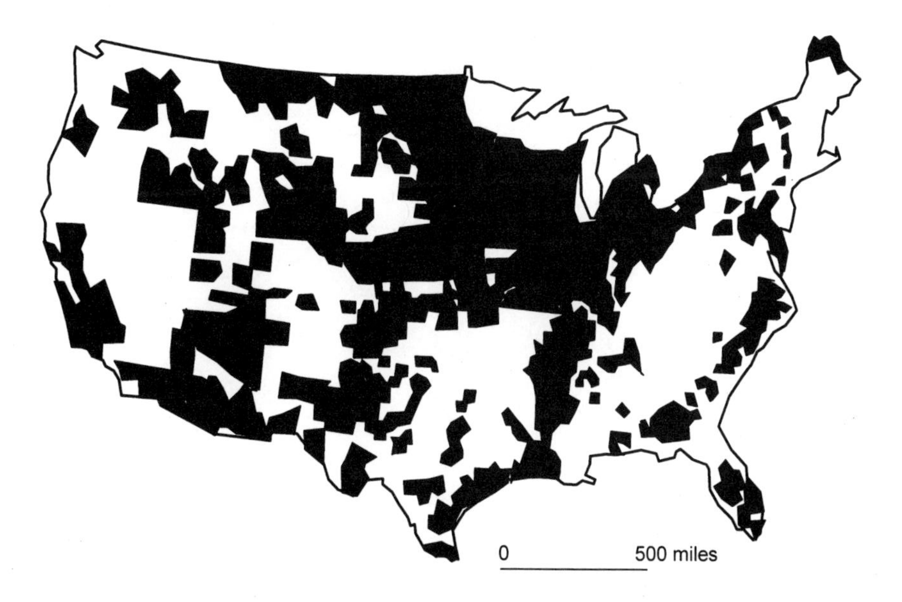

Areas of maximum annual estimated agricultural herbicide use in the United States by county, 1988–91. Non-crop use of herbicides is not included but can be substantial on suburban lawns and golf courses. (After P. Short and T. Colborn, 1999, Pesticide use in the U.S. and policy implications: A focus on herbicides, *Toxicology and Industrial Health*, 15, 240–75, Figure 3.)

led to reductions in the use of persistent pesticides such as DDT, which have been replaced by less persistent compounds.[42]

Water is one of the primary pathways by which these poisons move from their areas of application into the rest of the environment. Many of the pesticides are hydrophobic compounds, which means they are not very soluble in water. They are, however, highly soluble in fats, which allows them to accumulate in living organisms. They also have a strong tendency to sorb to organic material in soil and sediment, where they accumulate because they are resistant to degradation in the environment. Of the pesticides the NAWQA program tested for, 44 percent were detected in sediment and 64 percent in fish or mollusks. The organochlorine insecticides DDT, chlordane, and dieldrin were commonly detected in sediment or aquatic organisms at levels that may be toxic to aquatic life, wildlife, and people, even though their agricul-

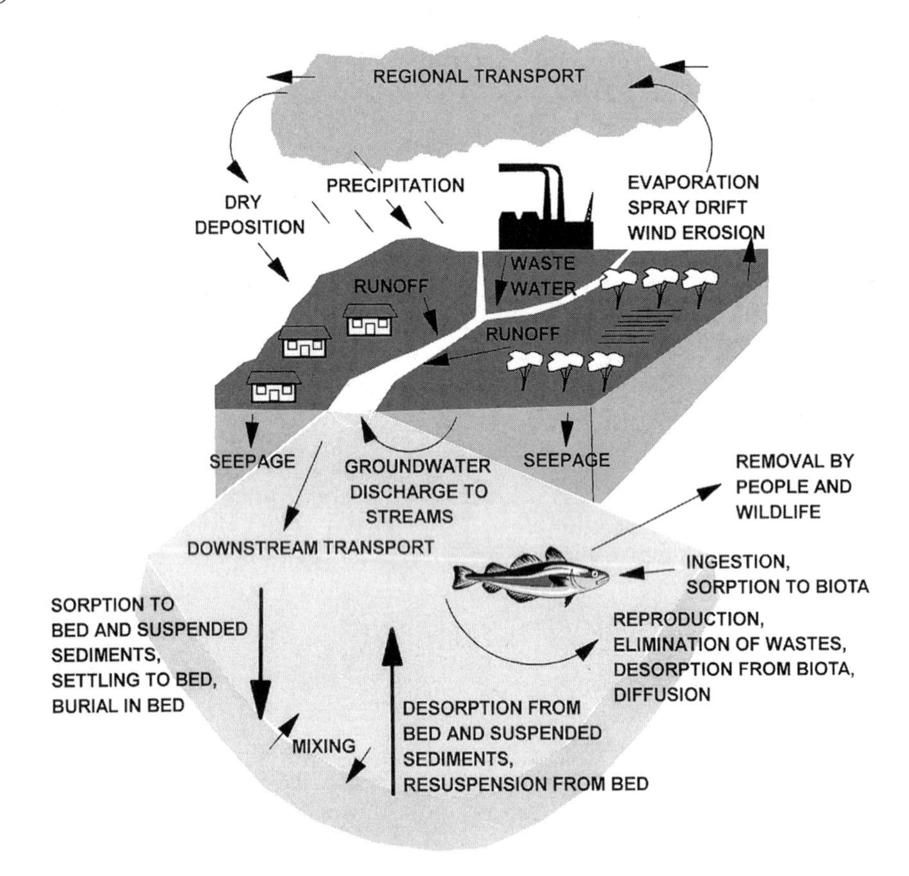

Pesticide movement through the water cycle, including air, soil, surface water and groundwater, and living organisms. (After M. S. Majewski and P. D. Capel, 1995, *Pesticides in the atmosphere: Distribution, trends, and governing factors*, Ann Arbor Press, Chelsea, Mich.)

tural uses in the United States have been banned since the 1970s. Although total DDT concentrations in fish have been declining since the 1960s, DDT still enters river ecosystems by atmospheric deposition and through erosion of previously contaminated soils. By 1990, DDT levels in the breast milk of women living in countries that banned DDT use during the 1970s fell below the standards for daily intake set by the World Health Organization, but these levels remained measurable.[43]

The highest rates of detection for the most heavily used herbicides—atrazine, metolachlor, alachlor, and cyanazine—were found in streams and shallow groundwater in agricultural areas. Insecticides were frequently detected in streams draining urban areas with high insecti-

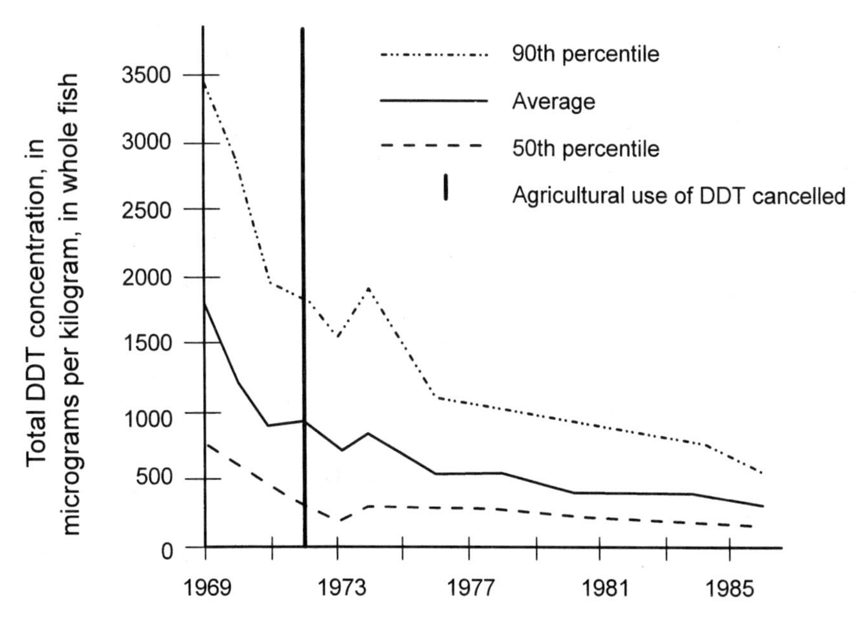

Trends in total DDT concentrations in whole fish sampled by the National Contaminant Biomonitoring Program from 1969 to 1986. (After U.S. Geological Survey, 2000, Pesticides in stream sediment and aquatic biota, *U.S. Geological Survey Fact Sheet FS-092-00*, Figure 5.)

cide use but were less frequently detected in shallow groundwater because most insecticides are applied in smaller amounts than herbicides and tend to attach to soil particles or to degrade quickly after application. Diazinon, carbaryl, chlorpyrifos and malathion were the most commonly detected insecticides. The greatest variety of pesticides occurred in rivers draining both agricultural and urban land. Rivers with poorly drained clayey soils, steep slopes, and sparse vegetation were most vulnerable to contamination, as were urban rivers, rivers draining tiled agricultural lands, or small rivers.[44]

Contaminated groundwater can be a major nonpoint contributor of pesticides to streams, and shallow groundwater less than one hundred feet below the surface in or adjacent to agricultural lands can contaminate domestic wells. No one monitors these domestic wells regularly, and homeowners in recently established residential areas are unlikely to be aware of persistent contamination.[45]

The atmosphere can also be a nonpoint source of pollution for streams. Nearly every pesticide investigated by NAWQA was detected

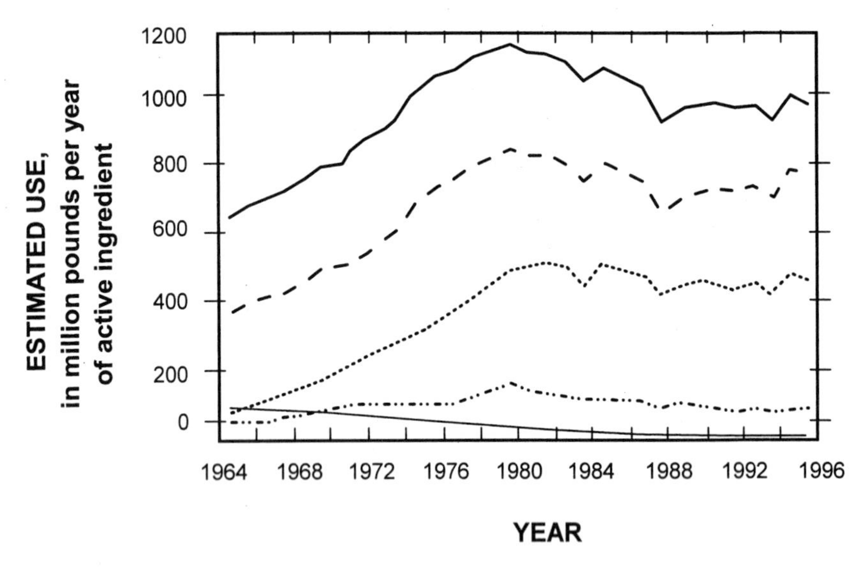

Changes in agricultural pesticide use, 1964–96. (After U.S. Geological Survey, 1999, The quality of our nation's waters: Nutrients and pesticides, *U.S. Geological Survey Circular* 1225.)

in air, rain, snow, or fog throughout the United States at different times of the year, with the highest concentrations occurring during the spring and summer months of high pesticide use. A 1999 circular published by the Geological Survey noted that "several instances have been recorded in which concentrations in rain have exceeded drinking-water standards for atrazine, alachlor, and 2,4-D." The simple pleasure of letting a spring or summer rain wash your face may be analogous to a toxic shower in areas downwind from agriculture. The circular also pointed out that "at least one pesticide was found in almost every water and fish sample collected from streams by NAWQA and in more than half of the shallow wells sampled in agricultural and urban areas. Moreover, individual pesticides seldom occurred alone." Concentrations of individual pesticides in samples from wells and as annual averages in streams were almost always lower than current EPA drinking-water standards and guidelines. But more than half of the agricultural and

Table 4.2. Summary Statistics for Detection Frequency (% of Sites Samples) of Organochlorine Compounds in River Ecosystems Under Various Types of Land Use

	Bed Sediment			Whole Fish			Clams		
Compound	Agricultural	Forest-Range	Urban	Agricultural	Forest-Range	Urban	Agricultural	Forest-Range	Urban
Total PCB	4.2	5.3	31.6	38.6	8.2	83.3	2.2	0	13.8
Total chlordane	8.8	2.7	60.7	38.6	8.2	79.7	8.5	3.8	55.2
Total DDT	49.6	22.4	70.5	87.7	38.4	91.7	34.0	7.7	31.0

Source: Data for 1992–98 from U.S. Geological Survey NAWQA Pesticide National Synthesis Web site, http://ca.water.usgs.gov/pnsp/oc_doc.html.

Herbicides

CYANAZINE

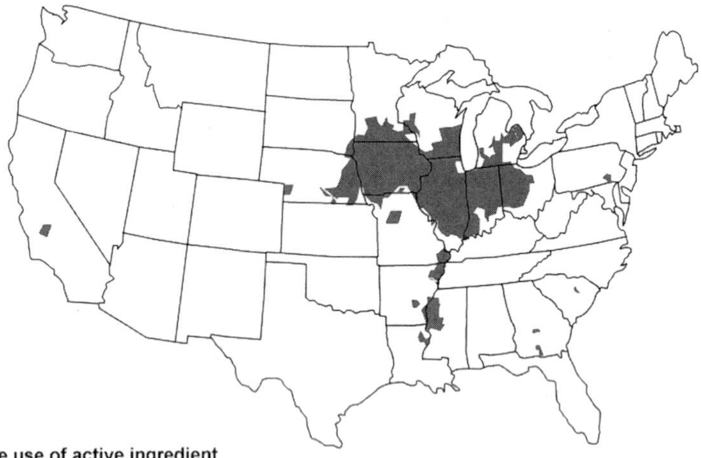

Average use of active ingredient
Pounds per square mile of county per year
Cyanazine > 32.7 [corn]
Metolachlor > 65.1 [corn]

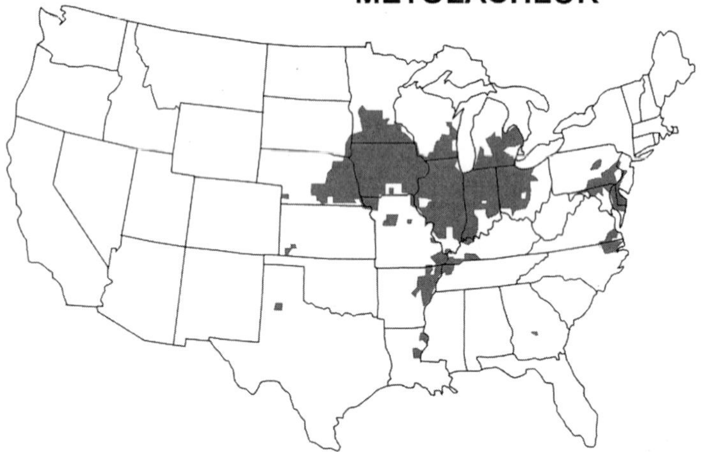

Maps of the highest national levels of the use of specified herbicides, by county, based on 1995 NAWQA data. The original, much more detailed maps on which these maps are based, as well as dozens of maps for other compounds, can be found at the NAWQA Web site: http://ca.water.usgs.gov/pnsp/use92.

Herbicides

Average use of active ingredient
Pounds per square mile of county per year
Atrazine > 66.5 [corn]
Alachlor > 29.2 [corn, soybeans]

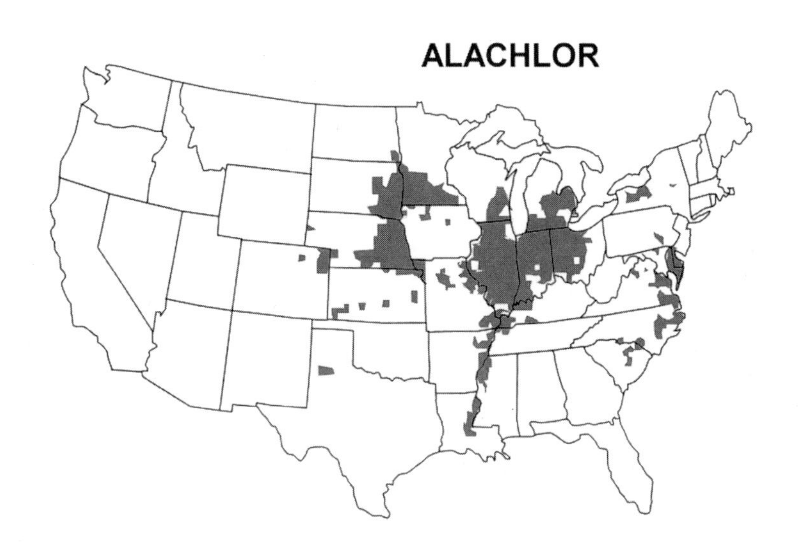

Insecticides

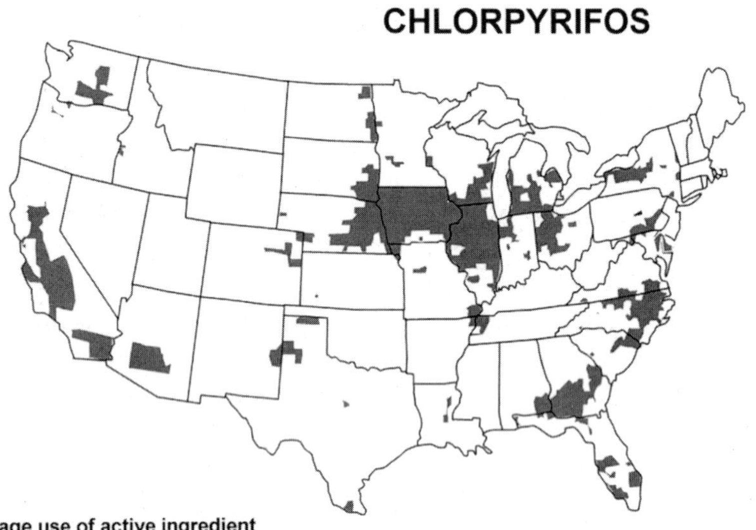

Average use of active ingredient
Pounds per square mile of county per year
Chlorpyrifos > 11.2 [corn, cotton]
Carbaryl > 3.3 [alfalfa, corn, pecans]

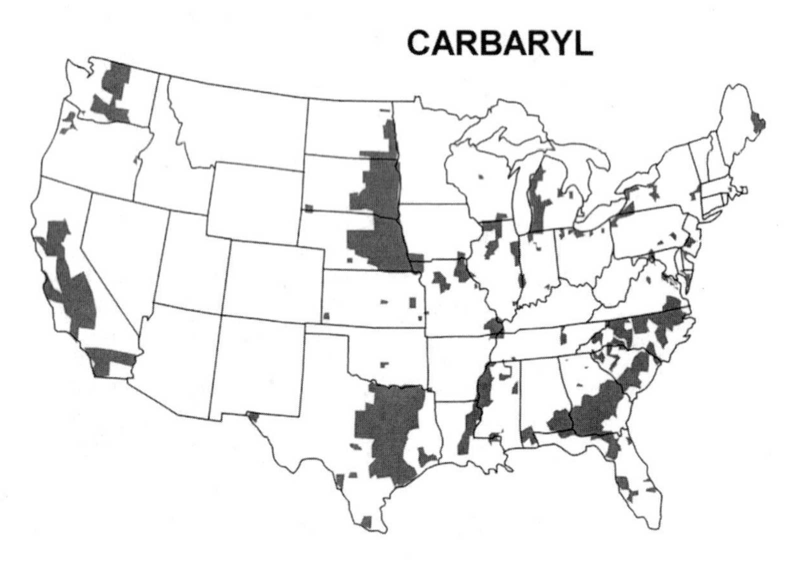

Maps of the highest national levels of the use of specified insecticides, by county, based on 1995 NAWQA data.

Insecticides

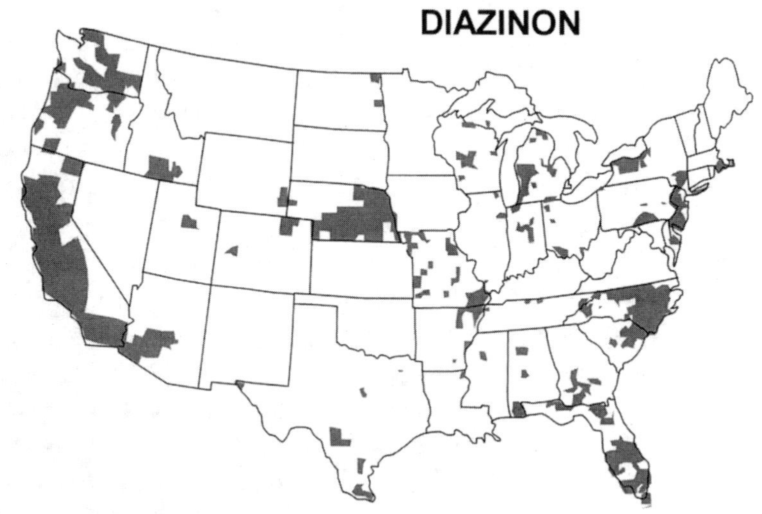

Average use of active ingredient
Pounds per square mile of county per year
Diazinon > 0.3 [almonds, plums, peaches]
Malathion > 1.2 [cotton, alfalfa]

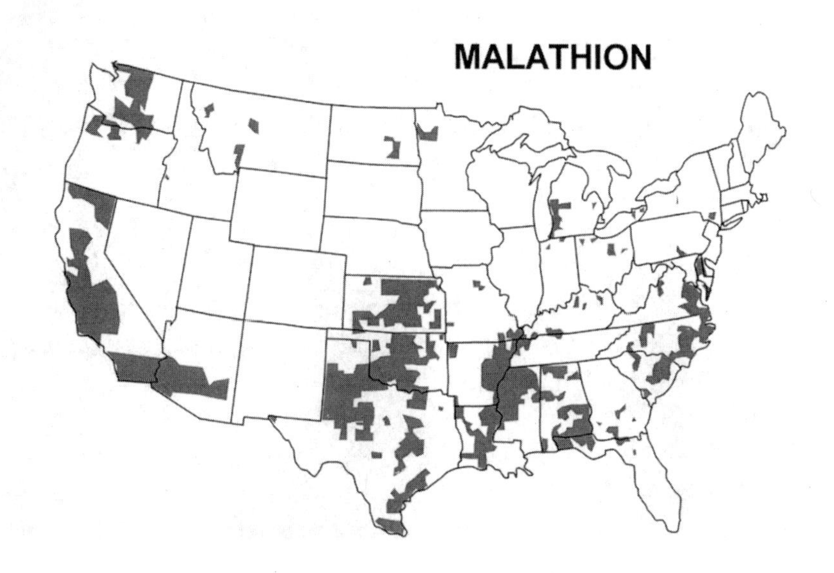

urban streams sampled had concentrations of at least one pesticide that exceeded guidelines for the protection of aquatic organisms. We cannot be complacent about even the drinking-water standards because of the critically important finding that pesticides are seldom present alone. As the Geological Survey noted: "Potential risks to humans and aquatic life implied by NAWQA pesticide findings can be only partially addressed by comparison to established standards and guidelines. Many pesticides and their breakdown products do not have standards or guidelines, and current standards and guidelines do not yet account for exposure to mixtures and seasonal pulses of high concentrations. In addition, potential effects on reproduction, nervous, and immune systems, as well as on chemically sensitive individuals, are not yet well understood. For example, some of the most frequently detected pesticides are suspected endocrine disrupters that have potential to affect reproduction or development of aquatic organisms or wildlife by interfering with natural hormones." In other words, we have drinking-water standards based on very simple, and mostly unrealistic, scenarios of only one contaminant being present in the water. We have no idea of what might constitute safe levels of multiple, combined contaminants.[46]

Prudence dictates that before a new toxic chemical is released for public use, it is tested thoroughly. How is the chemical transported within the environment once it is released? How long does it persist? What are its breakdown products? What does it combine with? Where does it accumulate? How does it affect a variety of living organisms throughout their life cycles when it is alone, combined with other compounds, or in breakdown products? But such tests are slow and costly. The tests are not done in the United States until after the new toxic compound is released, if they are done at all. Water-quality standards have been established for only about half of the pesticides measured in NAWQA water samples. Meanwhile, chemicals are cavalierly dumped together, or combined in surface and groundwaters draining many areas. These unintended chemical combinations very likely create new compounds that are more dangerous to living creatures than are the deliberately manufactured compounds, yet we have almost no data on these effects.[47]

Much of what we do know about these pesticides in the environment comes from biologists or biochemists working for government agencies, academic institutions, or nonprofit organizations. Ronald Eisler of the Fish and Wildlife Service synthesized existing reports on a variety of metals, pesticides, and other synthetic compounds and found that the behavior of these compounds in the environment varies widely, as seen when comparing atrazine and chlordane.[48]

The herbicide atrazine is the most heavily used agricultural pesticide in North America. More than 110 million pounds are applied annually to more than sixty-two million acres in the United States, primarily to control weeds in corn and sorghum. Atrazine degrades rapidly to less toxic compounds. After about four days, for example, half of the atrazine applied is gone from the soil, although it can persist for more than a year in dry, sandy soils in cool climates. Atrazine has limited bioaccumulation within organisms and biomagnification between organisms in aquatic ecosystems. The effects of atrazine on aquatic ecosystems are temporary, although residues of atrazine at toxic concentrations have been detected in groundwater, lakes, and streams fed by runoff from treated fields. Atrazine can affect aquatic animals at concentrations of 20 parts per billion; treated cornfields in Iowa produce runoff with concentrations of 4,900 parts per billion and sediments with 7,350 parts per billion shortly after application of atrazine. We understand little about the toxicity, environmental fate, and chemistry of atrazine and its breakdown products, and we are especially ignorant of its synergistic or additive effects with other agricultural chemicals in aquatic environments. But we do know that atrazine is linked to breast and ovarian cancer.[49]

Chlordane is an organochlorine compound introduced to the United States in 1947 as a broad-spectrum pesticide for use in homes, on lawns, and on agricultural and commercial properties. By 1974, approximately twenty-one million pounds of chlordane were produced annually. Concern over potential carcinogenicity led to a ban on chlordane in 1983, except for use against underground termites. Past chlordane use, coupled with atmospheric dissemination, has produced global contamination of the environment and of humans at low concentrations. The highest concentrations occur where chlordane is applied to termites. Half of the applied chlordane decays within about eighteen hours in water, although chlordane can persist more than a dozen years in soils. It is readily absorbed by warm-blooded animals through the skin, diet, and inhalation and is then distributed throughout the body. Chlordane does not have high rates of biomagnification in the food chain, but predatory species do have the highest concentrations, especially in the liver and body fat. Chlordane includes about forty-five components, and its breakdown product oxychlordane is much more toxic and persistent in mammals. Criteria for the protection of aquatic organisms are inadequate because we simply do not know enough about chlordane's effects on living organisms.[50]

Chlordane is one of a number of chemicals identified as endocrine disrupters. Endocrine glands such as the thyroid and pituitary secrete hormones into the blood or lymph. Hormones are the body's chemical messengers, regulating metabolism, reproduction, and mental processes, as well as prenatal development. Among the key hormones is estrogen. Estrogen and its target tissues interact in an intricate choreography as an organism develops. As estrogen bathes the fetus, the hormone's action is mediated through estrogen receptors on the cells. But other chemicals can bind to the estrogen receptors and disrupt the process, blocking communication between cells in the developing embryo. The timing of exposure to these chemicals is critical, for timing determines what stage of the embryo's development is most affected. In females, the organs most susceptible to developmental changes from these endocrine-disrupting chemicals include the mammary glands, fallopian tubes, uterus, cervix, and vagina. If exposure to these chemicals occurs during the prenatal or early postnatal periods, the effects are permanent and irreversible. Adults store these chemicals in their body fat, to be mobilized in some women from fat cells during pregnancy or lactation. There may not be any safe level of exposure to endocrine disrupters.[51]

Increasing public awareness of the effects of endocrine-disrupting chemicals comes from the work of scientists such as Theo Colborn, a zoologist with the World Wildlife Fund. In 1987, Colborn began to review scientific papers addressing the health of wildlife and humans in the Great Lakes region. When she began her review she was looking for evidence of increased cancer, the classic indicator of exposure to toxic substances. Instead, she noticed abnormalities in human and animal offspring, as well as in adults. Various studies now indicate a plethora of horrible effects across the full range of living organisms. Gastropods, fish, birds, and mammals in affected areas show signs of developmental disruption of secondary sexual characteristics, such as misshapen reproductive organs or abnormal reproductive behavior. Birds, fish, shellfish, and mammals have lowered fertility. Immune function is weakened in birds and mammals. Humans are subject to decreasing sperm counts, increasing testicular and breast cancer, and early female puberty, as well as memory deficiencies, abnormal sexual development, and behavioral changes such as attention deficit disorder. Increasingly, this array of ills is linked to the endocrine-disrupting chemicals released into the environment in large quantities since World War II.[52]

As Colborn has pointed out, people of the 1940s and 1950s formed the first generation to be exposed after birth to chemicals such as DDT and PCBs. Children of the 1950s through 1970s were the first generation with prenatal exposure, and during the period from the 1970s to the 1990s they have reached reproductive age. Case studies on this first generation of the brave new chemical world have shown direct evidence of the adverse effects of these chemicals. In 1977 the Michigan Chemical Company accidentally mixed the endocrine disrupter PBB into a batch of animal food. Cows, pigs, and chickens ate the contaminated food, and humans ate these animals before the error was discovered. A follow-up study indicated that girls whose mothers ate these animals experienced menarche an average of six months earlier than other girls. A twenty-year study of 594 children from North Carolina found that girls exposed to higher prenatal levels of PCB or DDE, as measured in the placenta, tended to be heavier and to mature earlier than girls with lower levels of prenatal exposure.[53]

Growing recognition of this contamination led to a multidisciplinary scientific conference in Wisconsin during 1991. Conference participants published the Wingspread Statement, an urgent warning that humans are being exposed to endocrine-disrupting chemicals demonstrated to disrupt development in both laboratory animals and wildlife populations. The statement goes on to say that unless these chemicals are controlled, we face the danger of widespread disruption of human embryonic development.[54]

The scientific community has provided evidence of endocrine disruption, particularly during embryonic development, for more than forty years. But the United States focuses on its existing protocols for testing new chemicals. These protocols were established to determine the probability of a chemical causing cancer or acute mortality, rather than to determine all potential adverse health effects. And many synthetic chemicals are not even tested. Of the approximately seventy-five thousand chemicals now in common commercial use, only an estimated

Table 4.3. Chemicals With Widespread Distribution in the Environment Reported to Have Reproductive and Endocrine-Disrupting Effects

Pesticides	Industrial Chemicals
Herbicides	Cadmium
2,4-D	Dioxin (2,3,7,8-TCDD)
2,4,5-T	Lead
Alachlor	Mercury
Amitrole	PBBs
Atrazine	PCBs
Metribuzin	Pentachlorophenol (PCP)
Nitrofen	Penta- to nonylphenol
Trifulralin	Phtalates
Fungicides	Styrenes
Benomyl	
Hexachlorobenzene	
Mancozeb	
Maneb	
Metiram complex	
Tributyl tin	
Zineb	
Ziram	
Insecticides	
β-HCH	
Carbaryl	
Chlordane	
Dicofol	
Dieldrin	
DDT and metabolites	
Endosulfan	
Heptachlor and H-epoxide	
Lindane (1-HCH)	
Methomyl	
Methoxychlor	
Mirex	
oxychlordane	
Parathion	
Synthetic pyrethroids	
Toxaphene	
Transnonachlor	
Nematocides	
Aldicarb	
DBCP	

Source: T. Colborn, 1994, The wildlife/human connection: Modernizing risk decisions, *Environmental Health Perspecives*, 102, 55–59, Table 1.

twelve hundred to fifteen hundred (1.5 to 3 percent) have been tested for carcinogenicity. We tend to be reactive rather than proactive with regard to problems of exposure to endocrine, nervous, and immune system disrupters. As a result, proof of causality is essential before regulatory action. The present approach of testing on a chemical-by-chemical basis in the laboratory rather than testing combinations of chemicals in an environmental setting makes it very difficult to establish such proof. Many of the endocrine-disrupting chemicals are capable of binding to intracellular estrogen receptors directly or after conversion to a breakdown product while in an organism or in the environment. Because of this, Colborn and others propose that chemical industries should be required to test all new products, and their breakdown products, both for multigenerational effects in at least three animal species and for their environmental fate in all media, including air, soil, and water.[55]

The NAWQA program produced several recommendations regarding pesticides. First, we need to know where the pesticides come from and how they move through the environment. Second, we need to address contamination within the framework of the entire hydrologic cycle acting across an entire drainage basin. For example, seasonal patterns of pesticide occurrence and concentration need to be characterized because these patterns dictate the timing of high concentrations in drinking-water supplies and the times when aquatic organisms may be exposed to high concentrations during critical stages of their life cycles. Third, we need to know how contaminants interact with each other and the environment, and how the contaminants affect all organisms throughout their life cycles. Current standards, guidelines, and associated monitoring programs do not account for contamination that occurs as mixtures of various parent compounds and degradation products. They also do not account for contamination that is characterized by lengthy periods of low concentrations punctuated by brief, seasonal periods of higher concentrations. Of the thousands of possible pesticide breakdown products, few have been looked for in streams or groundwater, even though some of the breakdown products have similar or greater toxicities than the parent compounds. Finally, we need to be able to predict pesticide occurrence, transport, and impacts associated with these poisons if we are to effectively reduce the hazards to aquatic ecosystems, and to ourselves.[56]

What is perhaps most amazing with respect to pesticide use is that it has not even accomplished its primary purpose. Total crop loss resulting from insect damage has doubled since synthetic pesticides were introduced for agricultural use at the end of World War II. Insect damage caused 7 percent of crop loss in the 1940s and 13 percent of crop loss at the end of the 1980s, but this statistic has been hidden by higher yields. Corn, one of our most common crops, had 3.5 percent loss to insects in 1945. Despite a thousandfold increase in insecticide use, 12 percent loss of corn was attributed to insects in 1997. Insects apparently evolve more rapidly than agents of chemical warfare.[57]

Nutrients in Groundwater and Surface Water

Nutrients are chemicals essential to plant and animal nutrition. The NAWQA program focused on nitrogen and phosphorus, nutrients that are important to aquatic life but can be contaminants at high concentrations. Nitrogen occurs in a variety of forms, including organic nitrogen and nitrate. Nitrate is a form of dissolved nitrogen highly soluble in water and readily transported in groundwater and streams because it is stable over a wide range of environmental conditions. Phosphates are a significant form of dissolved phosphorus in natural water. Phosphates are moderately soluble and not very mobile in soils and groundwater.[58]

Problems with nitrogen levels in agricultural soils date back nearly two centuries. Agricultural production declined in England and other parts of Europe by the 1840s, and famine might have occurred but for the discovery that the amount of nitrogen in the soil was the factor limiting food production. Nitrogen was added to the soil in the form of nitrate fertilizer, but by the end of the nineteenth century world demand was depleting the primary supply of nitrate deposits in Chile. A German chemist named Fritz Haber came to the rescue by discovering a method to economically mass-produce nitrogen. Haber and engineer Carl Bosch combined hydrogen and atmospheric nitrogen to produce ammonia, using uranium as a catalyst. Nearly 80 percent of the world's atmosphere is made up of extremely stable nitrogen molecules. Bacteria convert some atmospheric nitrogen first into ammonia and then into nitrites and nitrates, but these conversions do not occur at rates sufficient for modern agriculture. By accelerating the process of converting atmospheric nitrogen into a form chemically available to plants, Haber

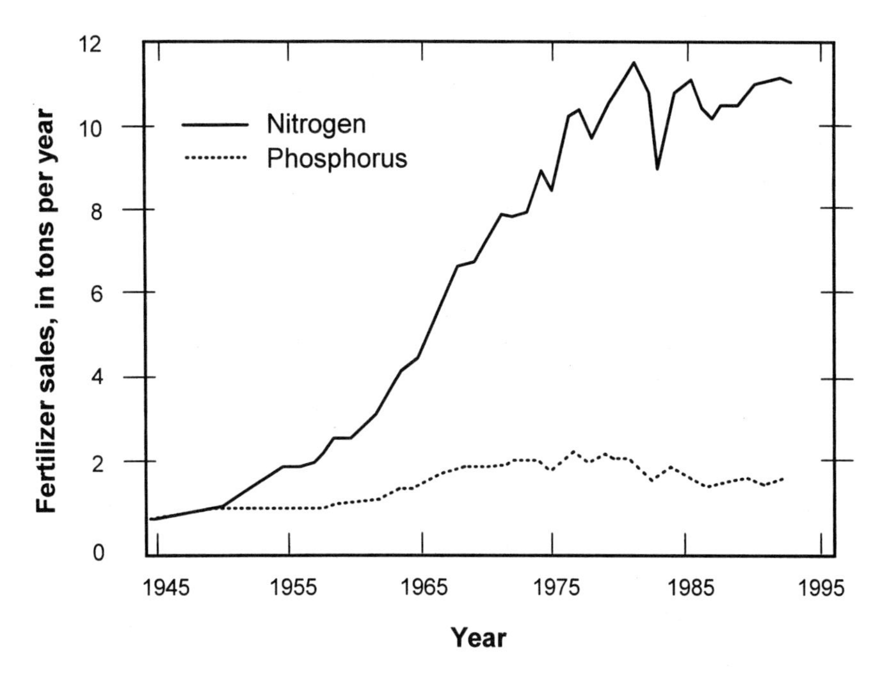

Annual sales of commercial nitrogen and phosphorus fertilizer in the United States. Between 1945 and 1985, the use of nitrogen fertilizer increased twentyfold; the use of phosphorus fertilizer increased about fourfold between 1945 and 1980. The annual use of both fertilizers remained relatively constant between 1989 and 1993. (After D. K. Mueller and D. R. Helsel, 1996, Nutrients in the nation's waters—too much of a good thing?, *U.S. Geological Survey Circular 1136.*)

and Bosch made possible the twentieth-century "Green Revolution" of agricultural production.[59]

Increased use of nitrogen and phosphorus fertilizers after World War II led to increased potential for water contamination by nutrients. From 1945 to 1993, the use of nitrogen fertilizer in the United States increased twentyfold, and the use of phosphorus fertilizer more than tripled. Inorganic fertilizer contributed about 9 million tons of nitrogen and about 1.8 million tons of phosphorus in the United States during 1993, with an additional 5.4 million tons of nitrogen and 1.8 million tons of phosphorus from animal manure. Atmospheric sources deposited about 2.7 million tons of nitrogen nationwide. Atmospheric sources result primarily from the burning of fossil fuels, which releases nitrogen oxides that are carried to surface waters as acid deposition or acid rain. Sources such as septic systems and leaking sewers contributed smaller,

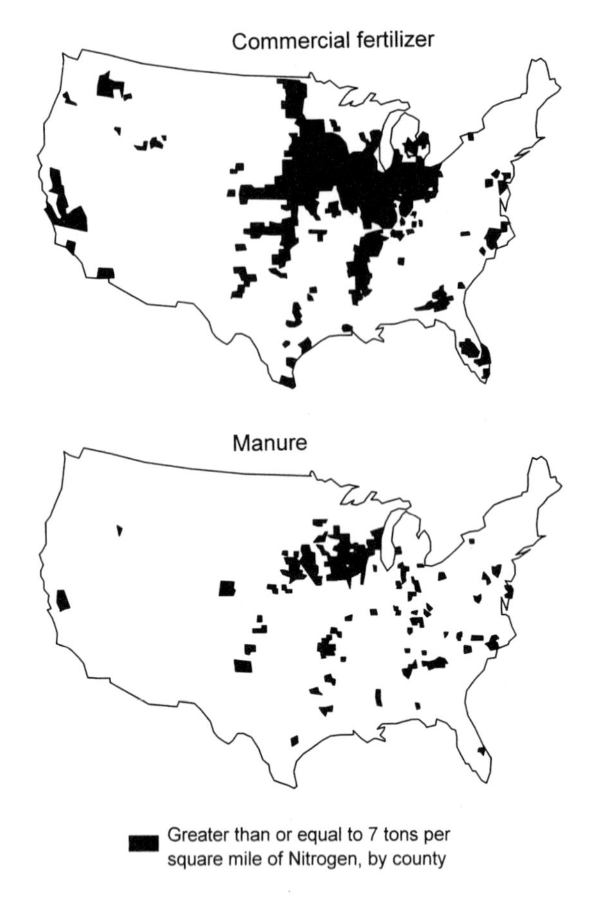

Estimated application of nitrogen in commercial fertilizer and manure during 1987. The highest application rates occurred over a broad area of the upper Midwest, with other high application areas along the East Coast, throughout the Southeast, and in isolated areas of the West. This figure likely underestimates the manure associated with confined animal feeding operations (massive feedlots and "factory farms"), which have become much more widespread since 1987. (After D. K. Mueller and D. R. Helsel, 1996, Nutrients in the nation's waters—too much of a good thing?, *U.S. Geological Survey Circular 1136*.)

but locally important, amounts of nitrogen. From about 1940 to 1970, laundry detergent was a major source of phosphorus to the environment. Contributions decreased substantially after state bans on phosphate detergents were enacted beginning in the 1970s. Agriculture is now the major contributor of phosphorus to streams.[60]

The NAWQA program detected nitrate in 71 percent of groundwater samples. Shallow groundwater within fifteen feet of the surface beneath agricultural land has the highest average concentration, followed by

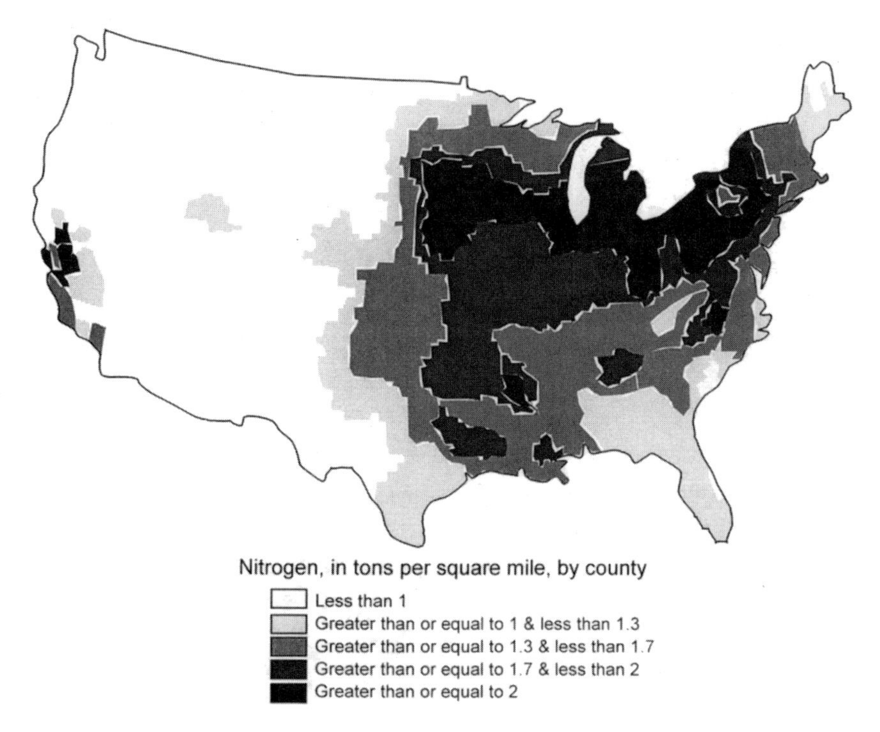

Estimated annual atmospheric deposition of nitrogen. The highest concentrations occur in a broad band from the Northeast through the upper Midwest. (After D. K. Mueller and D. R. Helsel, 1996, Nutrients in the nation's waters—too much of a good thing?, *U.S. Geological Survey Circular 1136*.)

shallow groundwater beneath urban land and deeper groundwater in major aquifers. Nitrate exceeded the maximum contaminant level of ten parts per million in more than 15 percent of groundwater samples from four of thirty-three major aquifers commonly used as a source of drinking water. Nitrate concentrations in groundwater are generally highest in parts of the Northeast, Midwest, and West Coast and lowest in the Southeast, as a result of differences in soil-drainage properties and agricultural practices. Nitrate in groundwater is highest under well-drained soils and intensive cultivation of row crops such as corn, cotton, or vegetables. Nitrate concentrations in surface water are generally lower than in groundwater but are elevated downstream from agricultural or urban areas. Elevated concentrations of nitrate in streams of the northeastern United States may be related to acid rain. Even alteration of headwater streams contributes more nitrogen. Small streams are sites of nitrogen uptake, and as these streams are destroyed by being piped into water-supply tunnels or filled with sediment, their ability to take up nitrogen is lost and more nitrogen is exported downstream.[61]

Contamination of groundwater by nitrate is a health concern because groundwater sources provide drinking water for more than half of the U.S. population. Ingestion of nitrate in drinking water can result in the "blue baby syndrome" of low oxygen levels in the blood, a potentially fatal condition. Excess nitrate in drinking water is also linked to spontaneous abortions and increased risk of non-Hodgkin's lymphoma, the most prevalent form of cancer in the United States. The EPA named nitrates and bacteria as the only contaminants that pose an immediate threat to health whenever base levels are exceeded.[62]

Elevated nitrogen and phosphorus concentrations in surface water are also of concern because they trigger eutrophication when excessive algal growth degrades water quality. Subsequent decay of the algae produces bad odors and taste and low levels of dissolved oxygen in the water. Algae produce oxygen when alive, but once they die, bacteria and other agents of decomposition use oxygen in the decomposition process and deplete dissolved oxygen levels in the water. Septic outflow into Lake Erie used to cause massive, stinking algal blooms that killed thousands of fish. Scientists link excessive nutrient concentrations in runoff from the Mississippi River to low dissolved oxygen in the Gulf of Mexico, and harm to fish and shellfish in the marine environment. The Gulf "dead zone" of low dissolved oxygen has in some years grown as large as New Jersey and extends from one to twenty yards down below the surface. The rich fishery of the Gulf is threatened by the same type of massive algal blooms that may have caused the collapse of the Baltic cod fishery in the early 1990s. Chesapeake Bay and the Albemarle-Pamlico Sound, the two largest estuaries in the United States, are also experiencing eutrophication. High nutrient concentrations may also contribute to the growth of the dinoflagellate *Pfiesteria* in Atlantic coastal waters. This form of algae is potentially toxic to fish, humans, and other organisms.[63]

Excessive algal growth contaminates drinking water in a manner more serious than unpleasant odor or taste. Greater phosphorus in agricultural runoff correlates with more algae in downstream reservoirs. More algae means more dissolved organic carbon, which in turn requires more intensive treatment to render the water drinkable. This treatment is expensive, and disinfection with chlorine produces byproducts such as trihalomethanes, which are linked to human bladder

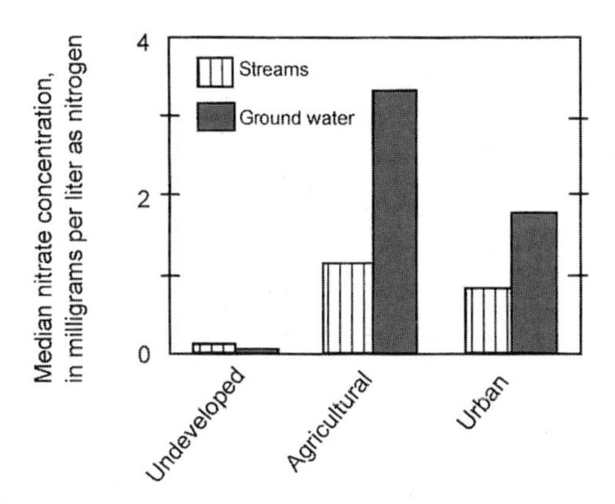

Nitrate concentrations in shallow ground water in agricultural areas were generally higher than in other areas. Concentrations in 12% of the domestic wells in agricultural areas exceeded the EPA maximum contaminant level for drinking water. Surface water rarely exceeded drinking water standards.

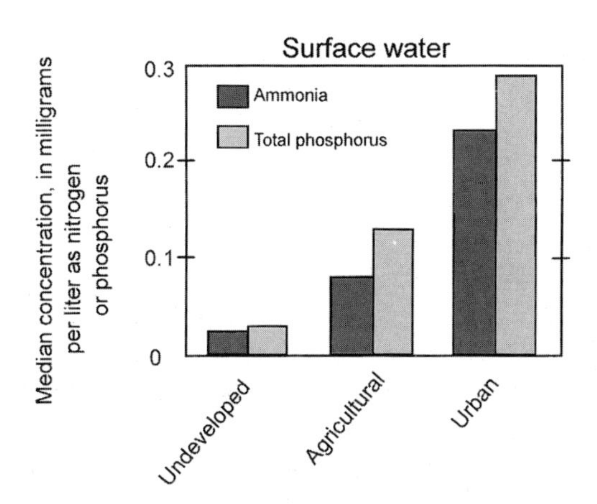

At least 10% of the surface water samples from urban sites contained sufficient ammonia to exceed the chronic exposure criteria for aquatic life. Total phosphorus concentrations exceeded the limit recommended by the EPA for streams in samples from urban and agricultural sites.

Concentrations of nitrogen and phosphorus in surface water and groundwater under differing land uses. (After D. R. Helsel and D. K. Mueller, 1996, Nutrients in the nation's waters: Identifying problems and progress, *U.S. Geological Survey Fact Sheet FS-218-96*.)

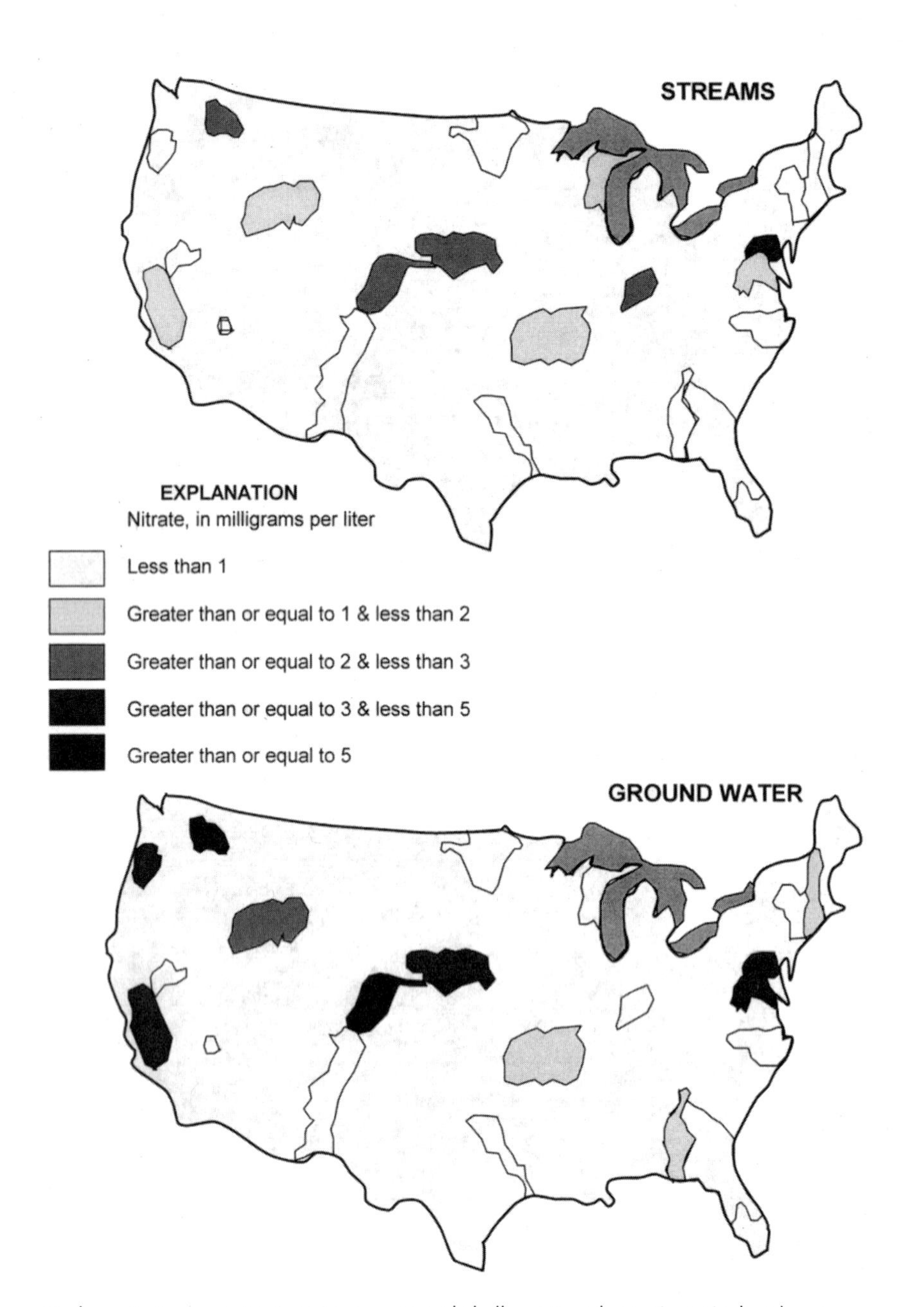

Median nitrate concentrations in streams and shallow groundwater in agricultural areas for twenty selected NAWQA study units. (After D. K. Mueller and D. R. Helsel, 1996, Nutrients in the nation's waters—too much of a good thing?, *U.S. Geological Survey Circular 1136.*)

cancer. National surveys by the EPA during the 1970s indicated the presence of trihalomethanes in virtually every chlorinated water supply tested. For people drinking chlorinated water, the rate of bladder cancer correlates with the amount of water consumed. As Sandra Steingraber writes in *Living Downstream,* "Giving people cancer in order to ensure them a water supply safe from disease-causing microbes is not necessary." Treatments such as granular activated charcoal and ozonation are immediate alternatives for drinking-water purification, but the EPA requires chlorine residual in the drinking-water distribution system.[64]

By the start of the twenty-first century, the planet's more than six billion humans produced more fixed nitrogen than did the soil's natural microbial processes. Both farmers and suburban users fertilizing lawns and golf courses generally apply more fertilizer rather than risk underapplication. This excess application occurs in part because commercial manufacturers of fertilizer tend to recommend unnecessarily high levels of application and in part because, knowing that nitrogen fertilizers are water soluble and thus partly washed off the site of application by rainfall, users overapply the fertilizers. Appropriate rates of application are very site-specific, depending on the plants to be fertilized, climatic conditions, and soil type. Relatively few counties or states have undertaken the studies necessary to develop detailed guidelines for fertilizer application. In the absence of such studies, excess fertilizer is commonly applied and more nitrogen accumulates than can be absorbed or broken down in the soil, supplemented by nitrates from fossil-fuel combustion. The excess, an estimated 20 percent of the nitrogen humans contribute to watersheds, ends up in freshwater systems and in the ocean. Increased nitrogen in water permits increased algal growth and eutrophication. The increased nitrogen alters aquatic ecosystems such as the Everglades, where species that evolved in a low-nutrient environment are displaced by exotic species more tolerant of nitrogen- and phosphorus-rich runoff. Increased nitrogen in soil also changes soil ecology, allowing more bacteria to produce the powerful greenhouse gas nitrous oxide, which is hundreds of times more effective than carbon dioxide at trapping heat within the Earth's atmosphere. At this point, however, we cannot feed ourselves without using nitrogen fertilizers. We must continue to develop methods to limit nitrogen emissions, better manage nitrogen fertilizer, and recycle nitrates via denitrifying bacteria.[65]

A NAWQA report summarized the scales at which we must address nutrient contamination: "No 'quick fixes' of long-term nutrient excesses should be expected. Ground water moves slowly, and waters of improved quality may take thirty years or more to move from the surface into nearby streams or wells. A long-term view must be taken. Understanding the regional distribution and key scientific factors that affect nutrient concentrations in ground and surface waters is critical to implementing and evaluating cost-effective programs to manage and protect our water resources."[66] Unfortunately, the combination of immediate changes in fertilizer use and associated costs and longer-term measurable responses in improved water quality and human and ecosystem health is a pairing unlikely to appeal to most politicians and voters.

Volatile Organic Compounds in Streams and Shallow Groundwater

Volatile organic compounds (VOCs) have chemical and physical properties that allow them to move freely between water and air phases. This mobility makes them widespread in the environment, although the specifics of their transport and storage are governed by several processes. Of these processes, volatilization—the movement of a VOC from water into the air—is the dominant control on the concentrations of VOCs in streams. Concentrations necessary to cause acute, or lethal, adverse effects on aquatic organisms are not likely to be routinely present in streams, although spills can result in short-term high concentrations. The effect of long-term chronic exposure to low concentrations of VOCs is largely unknown, but we are in the process of experimenting on ourselves. Many VOCs have properties that make them suspected or known health hazards to aquatic organisms and humans. To these substances we continually expose ourselves. Carcinogens such as the solvent tetrachloroethylene now flow with our tap water. Because this and other VOCs vaporize so readily, we absorb them across our skin and breathe them in as vapor. Hotter water produces more evaporation, and such apparently benign household activities as cooking or running a humidifier, dishwasher, or washing machine may release toxic chemicals into our bodies. Tests of VOC levels in the bloodstream indicate that ten minutes of a hot shower or thirty minutes of a hot bath produce a greater internal dose of VOCs than drinking half a gallon of water.[67]

The NAWQA program detected many of the same VOCs in both surface water and shallow groundwater from urban areas across the United States. Commonly detected VOCs included gasoline-related compounds such as toluene and chlorinated compounds such as chloroform. Urban land surfaces are the primary nonpoint source of most VOCs, with urban air as a secondary source. Flushing of spills and VOCs attached to organic materials, as well as leaking underground storage tanks, can also be important sources. The NAWQA program sampled 2,948 groundwater wells in the United States between 1985 and 1995. Forty-seven percent of the sampled wells in urban areas had at least one VOC, and 29 percent had two or more VOCs. Although concentrations were generally low, EPA drinking-water criteria were exceeded in 6.4 percent of all sampled wells and in 2.5 percent of the sampled drinking-water wells. In rural areas, 14 percent of the sampled wells had at least one VOC, and drinking-water criteria were exceeded in 1.3 percent of sampled drinking-water wells. The probability of finding VOCs in untreated groundwater is strongly correlated with population density.[68]

Methyl tert-butyl ether (MTBE), a petroleum product added to gasoline to abate air pollution or add octane, provides an example. The EPA has tentatively classified MTBE as a possible human carcinogen and issued a drinking-water advisory of twenty to forty parts per billion based on taste and odor. Although NAWQA did not sample groundwater known to be severely impacted by point-source contamination or runoff from regulated sites, NAWQA detected MTBE in 21 percent of 480 groundwater wells sampled in areas of the United States that use MTBE in gasoline. These areas are primarily the northeastern and mid-Atlantic states. MTBE was detected in only 2 percent of the wells elsewhere in the country. Most of these detections were below the EPA drinking-water advisory, but the frequency of detections above this advisory was ten times higher in the MTBE-use areas than in the rest of the nation. The Geological Survey estimates that up to twenty million of the fifty million people who obtain drinking water from groundwater in MTBE-use areas have a water supply that is vulnerable to contamination by VOCs. Vulnerability to contamination depends on the specific chemical's use, population density, and the presence of industry, commerce, and gasoline stations near the water supply.[69]

The Geological Survey's NAWQA program provides a unique perspective on contemporary water-quality conditions across the entire

United States. Because the program uses standardized procedures for sampling and analysis, it facilitates comparison among divergent geographic regions and land uses. Unfortunately, the next phase of the program, which began in 2001 and will continue through at least 2007, will be less comprehensive. Budget cuts forced the number of study units to be reduced from fifty-nine to forty-two. Work in the remaining forty-two units will focus on defining long-term trends in water quality and understanding the causes of these trends. Considering all of the indicators that the first phase of the program assessed, it becomes clear that every drainage basin in the United States has some deterioration of water quality as a result of human activities. Even forested watersheds with little direct human land use contain residues of such synthetic compounds as DDT or PCBs, which reached the watershed from atmospheric sources. Past actions to improve water quality have had positive effects. Reducing the use of phosphate detergents reduced excess phosphorus in water supplies. Integrated pest management, in which pesticides are selectively applied to areas with insect infestations rather than being automatically applied to all crops, has locally reduced organochlorine runoff. But room remains for substantial improvements in reducing trace elements, organochlorine compounds, nutrients, and VOCs entering surface water and groundwater from both urban and agricultural regions. These improvements must come from changes both in our use of these substances and in how we prevent excess or discarded substances from entering water supplies.[70]

Walking through my town, I watch stories unfold. A colony of ants erupts from a crack in the sidewalk and sprawl across the cement in an abstract pattern resembling a flower. The ants are still there the next day, and I notice a robin busily feeding on them. On the third day, a man with a spray can appears. He vigorously squirts the seething mass of ants, thoroughly saturating the area with poison. On the fourth day, the waters of a brief thundershower cleanse the now-lifeless sidewalk, flowing from the sidewalk down the storm sewer and into nearby Spring Creek. Some time later, I visit the post office. The surrounding lawns have been sprayed with pesticide and lined with the little yellow flags that do not explain to those who cannot read—children and animals—that the area is poisoned and off-limits. A toddler wanders across the lawn, plucking the gaily colored flags, and then presents them like a bouquet to her mother. An older child bicycling along the sidewalk takes a fall onto the grassy side rather than into the street. One in ten single-family households in the United States uses commercial lawn-care services; one in five applies the chemicals themselves. I see an abortion rights bumper sticker stating "Keep Your Laws Off My Body," and for a moment I misread it as "Keep Your Lawn Off My Body"—which is also an appropriate demand.

What is necessary to generate outrage over the poisons being freely and continually dumped into our air, soil, and water? Facts? Studies? We have them. Liver tumors in fish have been rising since 1940. In North America, liver tumors are now prevalent in sixteen fish species in at least twenty-five freshwater and saltwater sites. These tumors are virtually nonexistent in unpolluted waters. Pet dogs in households using 2,4-D are significantly more likely to have canine lymphoma than those in households without this toxin. Correlations are present between brain tumors in children and household use of pest-repelling strips, lindane-containing lice shampoos, flea collars, and lawn weed killers. Higher levels of DDT/DDE and PCBs in the blood correlate with breast cancer. Women younger than age forty who eat sport-caught fish from the Great Lakes nearly double their risk of developing breast cancer because these fish are contaminated with PCBs and DDT. Cancerous tissue contains high concentrations of DDE, PCBs, lindane, heptachlor, and dieldrin. Industrial countries have disproportionately more cancers, with adjustment for age and population size, than nonindustrial countries. The industrial countries contain 20 percent of the world's population, but 50 percent of cancer cases, and this cannot be attributed to better detection. *The World Health Organization concludes that at least 80 percent of all cancer is attributable to environmental influence.* We cannot blame genetics. The cancer rate of immigrant groups to industrial countries increases rapidly until it matches that of native-born citizens. And cancer among adoptees correlates within adoptive families, but not within biological families. Urine samples from people across the United States indicate that most of us have detectable levels of toxins such as chlorpyrifos, used in flea collars, in insect poisons, and on lawns and gardens. Death from cancer in the United States is highest along the northeast coast, the Great Lakes, and the mouth of the Mississippi River. These highest rates of cancer mortality correspond to the regions of the most intense industrial activity, but rates of cancer increase are greatest elsewhere, particularly in agricultural counties. More than 177 different organo-

Summaries of findings for individual drainage basins studied through the NAWQA program.

Hudson River Basin

* 13,400 sq mi; 55% forest, 33% ag, 7% urban
* 40 miles of river south of Ft. Edward is Superfund site because streambed sediments contain PCBs - exceed human health standards, among highest levels in US
* PCBs: significant environmental toxicity for invertebrates, fish, birds & mammals, & bioaccumulation in many organisms; consumption of contaminated food is major exposure route for humans
* Federal Energy Regulatory Commission developing remediation plan to remove upland soils & river sediments contaminated with PCBs
* Niagara Mohawk Power Corporation proposes to excavate soils & sediments; in 1923 NMPC's predecessor constructed Hudson River-Sherman Island development for hydroelectric power production, creating Sherman Island Reservoir (3.6 mi long & 305 acres). Reservoir required relocating the north shore road to a higher elevation. Areas between new road & reservoir leased as camps. Lessee of Queensbury camp site either disposed of PCB-laden oils onto surface of property, or used PCB-laden material for vegetation control &/or road oiling. 1992 investigation found PCB concentrations > 50 ppm, which constitutes hazardous waste.
* multiple sites along Hudson River have PCB contamination; concentrations highest in upper river reaches below Fort Edwards, and decrease with distance downstream
* PCB concentrations in streambed sediment strongly correlated with lipid-based PCB levels in fish tissues
* large-mouth bass collected at 3 locations on Hudson in 1994-95 indicate altered endocrine biomarkers, particularly in areas of elevated contamination in fish tissue & streambed sediments
* nutrient concentrations in urban & ag exceed those in forested areas
* highest yields for dissolved nitrate (> 3200 pounds/sq mi), total nitrogen (> 4200 pounds/sq mi), & total phosphorus (> 450 pounds/sq mi) from urban & ag; forest levels are < 1700 pounds/sq mi nitrate, <3400 pounds/sq mi total N, <150 pounds/sq mi phosphorus
* lowest suspended sediment in forests (0.06 tons/day/sq mi), highest in cleared watersheds (0.25 tons/day/sq mi)
* urban: chlordane
  streambed sediments have elevated metals, PCBs, PAHs; the most densely populated area, the Sawmill River at Yonkers, had some of the nation's highest concentrations of Cd, Cu, Hg, Ni & Zn in streambed sediments
  insecticides metolachlor, atrazine, deethylatrazine, diazinon, carbaryl, cyanazine
* DDT applied everywhere 1940-72, & detected everywhere

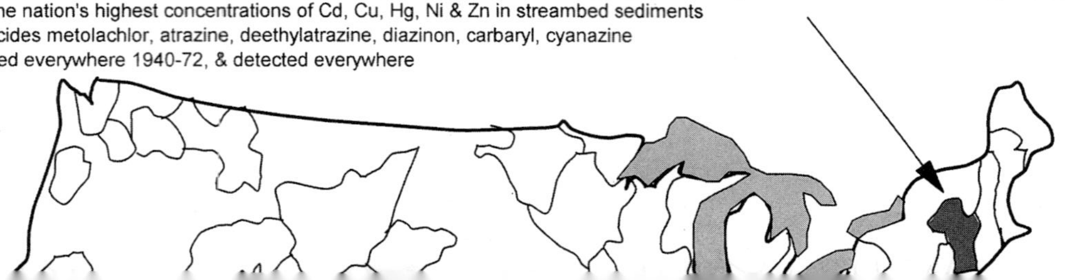

South Platte River Basin

* 24,300 sq mi; 41% rangeland, 37% ag, 16% forest, 3% urban (population 2.4 million)
* issues: in some areas, surface water standards not met in dissolved oxygen & un-ionized ammonia; nutrients; suspended sediment; ground water - nitrate exceeded drinking-water standards; water withdrawals
* in lower reaches of river, ground water is non-point source of nitrate, dissolved solids, atrazine & prometon; also, surface & ground water use have increased salinity
* urban & mixed use lands (urban & ag) have the highest levels of PCBs & organochlorines (DDT, dieldrin) in streambed sediment & fish tissue
* streams with historical mining have elevated trace metals
* VOCs in 86% of shallow urban ground water samples, & even present in some ground water samples from forested areas
* water quality is generally good in forested mountain sites
* ag areas have most degraded water quality (nitrates & salinity in ground water, salinity & suspended sediment in surface water)
* water quality in urban areas degraded with PCBs, VOCs, organochlorine pesticides

Nevada Basin and Range

* 4,000 sq mi; 60% range, 4% urban (population 1.1 million), 2% irrigated ag
* urban

N, 24 pesticides (simazine, prometon, diuron, DCPA, diazinon & malathion)
aquatic life criteria for diazinon & malathion exceeded in 47% & 25% of samples, respectively
pesticides in some ground water samples
VOCs in surface & ground water
Cd, Cr, Cu, Pb & Zn enriched
evidence of disruption of carp endocrine systems in Las Vegas Wash & Las Vegas Bay of Lake Mead

* ag

dissolved solids, As, boron,& molybdenum commonly exceed criteria for protection of aquatic life -- leached from desert soils by irrigation
organophosphate & nitrate in ground water
pesticides atrazine, simazine & prometon in surface & ground water

* historical mining has caused Al, As, Cr, Cu, Pb, Hg, Ag & Ni in streambed sediments

Upper Illinois River Basin

* 11,000 sq mi; 75% ag; population 7.6 million, with 6 million in urban areas (including Chicago)
* issues: atmospheric deposition of pesticides & trace metals; endocrine disrupters in surface & ground water; nutrients; aquifer level recovery; urbanization
* ag: diverse fish communities, with species intolerant of contaminants

pesticides alachlor, atrazine, cyanazine, metolachlor, metribuzin

* urban: fewer, more tolerant fish species

contaminant concentrations rarely exceed standards for protection of aquatic life
pesticides bromacil, diazinon, malathion, prometon, simazine

* Chicago: water, sediment & fish tissue have elevated trace elements (Al, As, Ba, Be, Bo, Cd, Cr, Co, Cu, Fe, Pb, Mn, Hg, Mo, Ni, P, Se, Ag, Sr, Vn, Zn) - some ag areas also have elevated concentrations

tetrachloroethylene, trichloroethylene & 1,2-dichloroethylene exceed maximum contaminant levels for drinking water

* nutrient loads & yields among highest in the Mississippi River drainage basin; 1978-97, 91,800 tons/year of N export & 300,000 tons/year N input to streams from ag, urban & other sources
* DDT, chlordane, dieldrin common
* high N, high bacteria, low dissolved oxygen
* strong correlation between fish community, water quality, & Cr in streambed sediments
* atrazine exceeds maximum contaminant levels for drinking-water during runoff in some areas
* all stations with at least 1 VOC were within 2 miles downstream of a point source

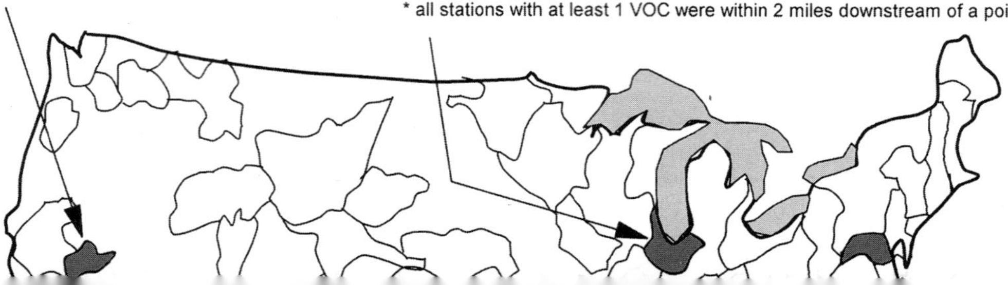

Trinity River Basin

* 18,570 sq mi; 57% ag, 25% forest,
  5% urban (population 4.5 million)
* 90% surface water use
* urban

chlordane, DDT, PCBs, dieldrin in fish

chlordane, PAHs & Zn concentrations in streambed sediments increase since mid-1960s

Pb, DDT, PCBs concentrations in sediments decrease with time

2-4 insecticides present; diazinon exceeds drinking-water maximum contaminant levels in 15% of samples

* general

4-6 herbicides present

atrazine ubiquitous, often exceeds drinking-water maximum contaminant levels

Lower Susquehanna River Basin

* 9,200 sq mi; > 50% forest, 35% ag, 15% urban (1.85 million population)
* issues: nutrient & sediment loads; pesticides; acid precipitation; acid-mine drainage & industrial waste; radiochemicals
* 30% of ground water samples & 20% of stream samples exceed EPA maximum contaminant levels for nitrate-n (from animal manure as ag fertilizer)
* most common pesticides are herbicides atrazine, metolachlor, simazine, prometon, alachlor & cyanazine -- mostly below drinking-water standards
* coliform bacteria in 70% of household wells
* VOCs in Harrisburg wells, but below maximum contaminant levels
* mine drainage: As, Be, Cd, Co, Fe, Mn, Ni, Se & Zn
* PCBs, DDT, & chlordane in streambed sediments & fish tissue
* radon in ground water

Upper Snake River Basin

* 35,800 sq mi; 50% forest & grazing land, 33% irrig ag & ground water recharge
* issues: ag; grazing; hydroelectric dams; source of mercury found in fish tissue
* Middle Snake River from Milner Dam to King Hill is water-quality limited under Clean Water Act because of excessive aquatic vegetation, low dissolved oxygen, & high temperature -- eutrophic because of excessive nutrient & sediment inputs and decreased stream flows
* nutrients from ground water discharge, fish hatcheries, municipal discharge & irrigation returns

Red River of the North Basin

* 35,530 sq mi; 74% ag, 26% forest, 3% urban, population 418,000
* extensive pesticide use, but small amounts in streams
* 2,4-D, MCPA, bromoxynil & trifluralin most heavily applied (also atrazine, cyanazine, metolachlor found in more than 50% of samples)
* 253,000 pounds of atrazine applied annually; 1,340 pounds (0.5%) left basin via streams during wet year of 1993
* Cd, Pb, Hg, PCBs in streambed sediments & fish
* DDT, PAHs in streambed sediments - low concentrations, but of concern
* locally problems with N, P & suspended sediment; streams with lakes, reservoirs & wetlands have lower sediment concentrations & yields

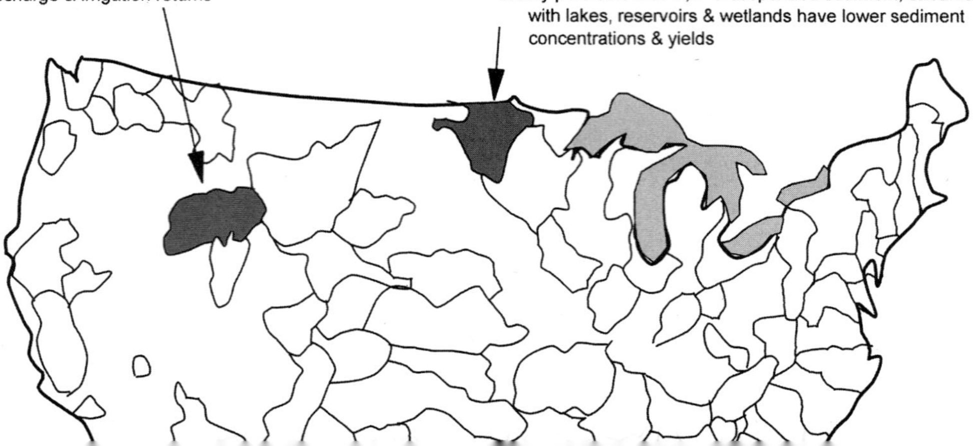

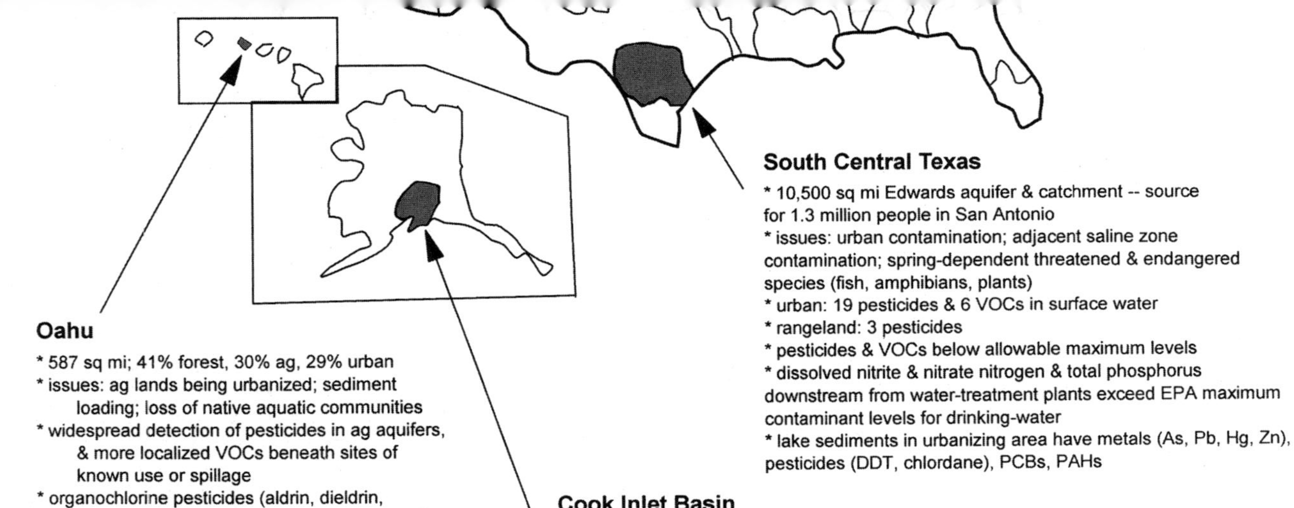

South Central Texas

* 10,500 sq mi Edwards aquifer & catchment -- source for 1.3 million people in San Antonio
* issues: urban contamination; adjacent saline zone contamination; spring-dependent threatened & endangered species (fish, amphibians, plants)
* urban: 19 pesticides & 6 VOCs in surface water
* rangeland: 3 pesticides
* pesticides & VOCs below allowable maximum levels
* dissolved nitrite & nitrate nitrogen & total phosphorus downstream from water-treatment plants exceed EPA maximum contaminant levels for drinking-water
* lake sediments in urbanizing area have metals (As, Pb, Hg, Zn), pesticides (DDT, chlordane), PCBs, PAHs

Oahu

* 587 sq mi; 41% forest, 30% ag, 29% urban
* issues: ag lands being urbanized; sediment loading; loss of native aquatic communities
* widespread detection of pesticides in ag aquifers, & more localized VOCs beneath sites of known use or spillage
* organochlorine pesticides (aldrin, dieldrin, chlordane, DDT) widespread in streambed sediments & fish tissue in urban, ag, & mixed land use areas -- used against termites -- endocrine-disrupters

Cook Inlet Basin

* 39,300 sq mi; 30% forest, 20% glaciers, 12% lakes & wetlands, < 1% ag, < 1% urban
* issues: salmon spawning; documenting conditions in undisturbed environments
* found no contaminants in streams in national parks & wildlife refuges
* DDT, PCBs in some drainages, up to 23 VOCs at other sites -- highly correlated with population density
* As, Cd, Cr, Cu, Pb, Ni, Se, Zn in streambed sediments

Puget Sound Basin

* 13,600 sq mi; 86% forest, 6% ag, 9% urban or rural (3.4 million population)
* issues:
  surface water -- loss of habitat, urban runoff, nutrients
  ground water -- pesticides, nitrate
* ag lands: herbicides atrazine, prometon, simazine & tebuthiuron in surface and ground water
* urban lands: insecticides carbaryl, diazinon & malathion, & herbicides
* pesticides often exceeded standards for protection of aquatic life, but not drinking-water standards

Connecticut, Housatonic and Thames River Basins

* 16,000 sq mi; 78% forest, 12% ag, 10% urban
* concentrations of PCBs in streambed sediment and fish tissue in Housatonic among highest in nation
* nitrogen and phosphorus in surface water
* most frequently detected:
  ag: herbicides atrazine, metolachlor and simazine
  urban: herbicides prometon; insecticides diazinon and carbaryl
* drinking water standards not exceeded, actual health concern unknown
* groundwater: 24 pesticides, VOCs, nitrate
* sediments: Cr, Cu, Pb, Hg, Ni, Zn, chlordane, DDT, PAHs, PCBs throughout basin - lowest in northern forested basins, highest in southern urbanized basins

San Joaquin-Tulare Basins

* 31,200 sq mi; forest & ag
* detected 49 pesticides, 7 at concentrations exceeding criteria for protection of aquatic life, none exceeding drinking-water standards
* most frequent pesticide occurrence & highest concentrations correspond to heaviest ag applications
* diazinon coming from dormant orchards
* DDT, toxaphene & chlordane in streambed sediment & tissues of clams & fish (bioaccumulation)
* DDT, chlordane & dieldrin exceed concentrations for protection of aquatic life
* nitrate concentrations have increased in last 40 years, but still below drinking-water standards
* introduced fish species outnumber native species almost 2:1
* groundwater: nitrate concentrations exceed drinking-water standards; low pesticide concentrations in 2/3 of samples from domestic water supply wells
* in 1990, ag used 14.7 million ac-ft of water and 597 million pounds of nitrogen & phosphorus fertilizers; in 1991, ag used 88 million pounds of pesticides
* in 1987, livestock contributed 318 million pounds of N & P from manure
* almost every major river from Sierras has 1 or more reservoirs; almost every tributary and drainage altered by network of canals, drains and wasteways

Georgia-Florida Coastal Plain

* 62,000 sq mi; 48% forest, 28% ag, 9.3 million population
* mostly use ground water
* nitrate exceeds maximum contaminant levels for drinking-water in 20% of ground water in ag areas
* N, P, & 21 pesticides in ground water, 32 pesticides in surface water
* pesticides don't exceed drinking-water standards, but sometimes exceed standards for protection of aquatic life
* most common pesticides are atrazine, metolachlor & prometon
* groundwater: 11 VOCs, high radon (natural)
* organochlorine pesticides exceed standards for protection of aquatic life in 22% of streambed sediment samples - - mostly chlordane & DDT
* PCBs in urban streambed sediments

chlorine residues can be detected in the body of an average middle-aged man in the United States.[71]

This is our body burden, the sum total of our exposure to synthetic chemicals via all routes of entry and all sources. If these chemicals are soluble in our fat, and persistent, then we develop a cumulative body burden that increases as we age. So why is there no collective outrage over what can only be termed an epidemic of cancer in the industrialized world? One reason is the lack of large studies attempting to link the incidence of cancer to environmental contamination. All of the studies summarized in the previous paragraph were relatively small in scope. The lack of large studies results from a lack of funding; the government has never appropriated money for such studies. Peter Infante, the director of the Health Standards Program at the Occupational Safety and Health Administration (OSHA), is quoted in Sandra Steingraber's *Living Downstream:* "'We need more study' is the grandfather of all arguments for taking no action." Yet such studies are not being done. Steingraber, using Illinois as an example, describes how a state cancer registry was funded to track cancer occurrence and mortality. A state health and hazardous substances registry act, proposed simultaneously to collect information on exposure to hazardous substances, was not funded. The 1994 National Cancer Advisory Board's report to Congress emphasized that a lack of appreciation for environmental and food-source contaminants has frustrated cancer prevention efforts, yet physicians and drug companies continue to emphasize the role of genes and heredity and lifestyle risks such as smoking.[72]

There is no question that lifestyle choices such as smoking dramatically increase an individual's risk of cancer. But we should not become so obsessed with these trees that we miss the forest. Steingraber describes in detail the cancer and reproductive failure destroying the beluga whales of the St. Lawrence River estuary. Their fat has high levels of PCBs, DDT, chlordane, and toxaphene. PCBs and DDT were manufactured and used in the drainage basin upstream from the estuary. The other contaminants were blown in on the winds, or carried in the tissues of eels migrating from the Sargasso Sea and the Ontario basin. The whales eat these eels, ingesting toxins with each mouthful. As Leone Pippard of Canadian Ecology Advocates demanded: "Tell me, does the St. Lawrence beluga drink too much alcohol and does the St. Lawrence beluga smoke too much and does the St. Lawrence beluga have a bad diet . . . is that why the beluga whales are ill? . . . Do you think you are somehow immune and that it is only the beluga whale that is being affected?"[73] Underground miners in past centuries took a caged canary with them into the mines. If the canary died, the miners had only moments to flee the site in response to this sign of poisonous gas. The St. Lawrence belugas and hundreds of other species are our contemporary dying canaries. We continue to ignore them at our own peril.

The Great Lakes Region Today

The rivers of the Great Lakes region remain heavily polluted in many areas despite the reductions in sewage input. The pollutants recognized during the late twentieth and early twenty-first centuries have accumulated since the start of industrialization in the case of heavy metals, and since the 1940s in the case of organochlorine pesticides, PCBs, and VOCs. These toxins are not as readily flushed from a river system as are fermentable organic wastes, and they persist at deadly concentrations for decades after the source of contamination to the river has ceased. The upper Illinois River basin no longer receives untreated Chicago sewage, but it has heavy metal concentrations reflecting industrial and municipal activities. High concentrations of arsenic occur along the tributary Kankakee River. Cadmium, chromium, copper, lead, mercury, nickel, silver, and zinc are present along the Des Plaines and Illinois Rivers. Strontium contaminates the Fox River, where some of the first U.S. studies of environmental carcinogens identified fish tumors resulting from heavy metals leaching out of old industrial sites and strip mines. The Fox River empties into Green Bay in Lake Michigan. The river transports 70 percent of the sediment reaching Green Bay, and that sediment is contaminated with nutrients and PCBs. Despite the simple lake geometry around the river mouth, storms and waves redistribute the contaminated sediment in complex patterns that are hard to map or control.[74]

An international commission on the Great Lakes identified forty-two sites of concern where sediments are contaminated by PCBs, PAHs, dioxins, and metals. More than 20 percent of the total shoreline of the lakes is impaired. Ninety-nine percent of these two thousand miles are

The Sheboygan River, Wisconsin, downstream from the Kohler Company landfill.

impacted by toxic organic chemicals. Metal contamination is ubiquitous. The Sheboygan River and Harbor Superfund site is one of five areas of priority concern in the Great Lakes and provides an example of the history and implications of toxic contamination.[75]

The Sheboygan River drains 260 square miles of Wisconsin, emptying into Lake Michigan. PCBs contaminate the lower fourteen miles of the river and its floodplain. Fish sampling in 1975 and 1976 did not indicate any contamination of the river, but sampling in 1977 showed high levels of PCBs—two hundred parts per million—in the tissues of fish living in the lower river. Advisories were issued to fishers in 1978 (what are commonly referred to as "fish advisories" do not, of course, advise the fish to move elsewhere), and basin-wide sampling and monitoring began. Three industries along the river were identified as the sources of PCB contamination. Thomas Industries, an aluminum die cast shop that cleaned aluminum parts with degreasing products, admitted responsibility and began to work with the Wisconsin Department of Natural Resources on remediation and cleanup. Tecumseh Products Company and the Kohler Company still contested their responsibility as of 2002.[76]

Tecumseh operates an aluminum die cast plant making engine blocks for lawnmowers and other gasoline engines, power generators, and refrigeration and air-conditioner compressors. Tecumseh added PCB hydraulic fluid to the oil in the hydraulic presses during the 1970s, and then dumped the hydraulic fluid behind the plant into a bermed pond that eroded into the Sheboygan River. EPA put the site on the National Priority List in 1986, and during 1989–90 an emergency removal of forty-three hundred cubic yards of contaminated sediment was conducted. Sediment removal and remediation continue at the site.[77]

Kohler Company is the second largest plumbing manufacturer in the nation. Despite concentration gradients of contaminants that clearly indicate PCBs leaching from the Kohler landfill, the company refuses to admit responsibility or take any action. As with other Superfund sites, more money is being spent on identifying the responsible parties than on remedying the situation. (The Superfund act is sometimes called the Lawyers Full Employment Act.) In 1997 a groundwater-interceptor drain was installed below the Kohler landfill to limit the spread of contaminants. Maximum PCB levels (greater than fifty parts per million PCBs) are now found six to thirteen feet below the surface of the river and harbor sediment because the recently deposited sediments are less contaminated. Sediment removal and remediation are scheduled to begin in 2004, with approximately twenty-one thousand cubic yards to be removed from the upper river and fifty thousand cubic yards from the lower river and inner harbor at an estimated cost of $41 million. (These volumes are approximately equal to 3.3 and 8 football fields, respectively, covered in sediment one-yard deep.) The contaminated sediment will be put in storage facilities and then transported by train to Oklahoma. As with nuclear waste, we as a nation tend to ship our toxic wastes to the states impoverished, underpopulated, politically corrupt, or ignorant enough to accept them. However, the whole process of cleanup at the Sheboygan site may not occur after the Bush administration's 2002 budget cuts that crippled the Superfund program.[78]

The Lake Erie–Lake St. Clair basin, which includes the Cuyahoga River, is one of the NAWQA study units. As of 1990, the 22,300 square miles of the drainage basin contained 10.4 million people. Seventy-five percent of the land in the basin is agricultural, 11 percent is urban, and 10 percent is forested. The amounts of nitrogen and phosphorus fertilizers, pesticides, and sediment discharged to Lake Erie from its drainage basin are higher than the amounts discharged from any other basin of the Great Lakes. More than thirty years after the Cuyahoga River burned,

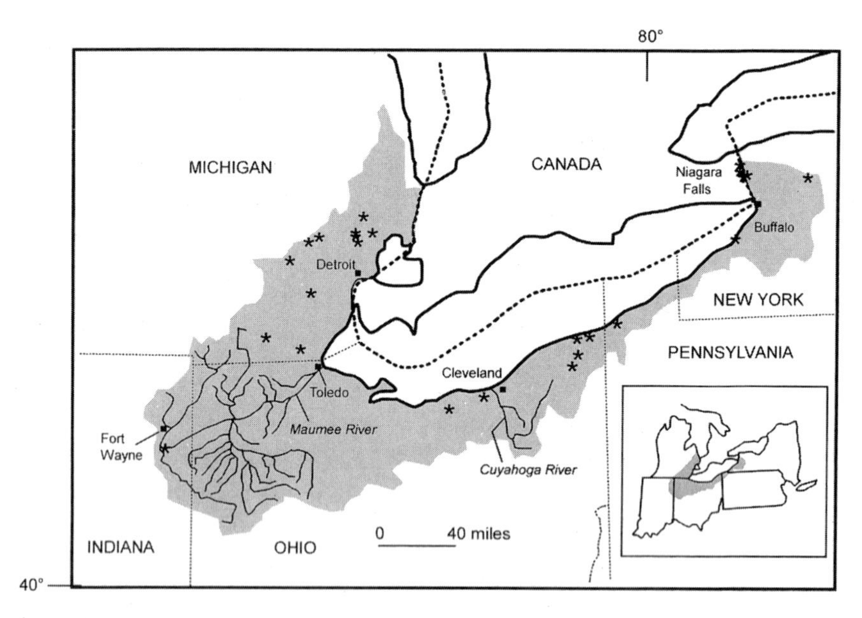

The Lake Erie–Lake St. Clair basin NAWQA study unit (shaded). Asterisks indicate National Priority List abandoned toxic-waste sites that pose a potential risk to life or health. (After G. D. Casey, D. N. Myers, D. P. Finnegan, and M. E. Wieczorek, 1998, National water-quality assessment of the Lake Erie-Lake St. Clair Basin, Michigan, Indiana, Ohio, Pennsylvania and New York—environmental and hydrologic setting, *U.S. Geological Survey Water-Resources Investigations Report 97-4256*.)

the Lake Erie drainage basin still contains many of the nation's toxic hot spots. Toxic Release Inventory (TRI) sites are those that process at least twenty-five thousand pounds of listed toxic chemicals per year. The Lake Erie basin has 2,438 TRI sites, which process more than three hundred different types of toxic chemicals in twenty different chemical categories. The basin also contains thirty National Priority List abandoned toxic-waste sites that pose a potential risk to life or health.[79]

Urban impacts to rivers in the Lake Erie basin include nutrient enrichment and oxygen depletion. Nitrogen, phosphorus, and fecal bacteria all reach high levels in some portions of the basin. The drainage basin has a wet climate; between thirty and fifty inches of precipitation fall each year on average, producing from eight to twenty inches of annual runoff into the streams. This runoff may be contaminated with *E. coli* and fecal coliform bacteria from septic systems. More insidious are the PCBs, PAHs, and trace elements—arsenic, cadmium, lead, zinc, chromium, and copper—entering the rivers from urban areas. PAHs are polycyclic aromatic hydrocarbons, which are by-products of combustion produced when fossil fuels are consumed to produce energy. Like the other contaminants, PAHs are carcinogenic and do nothing good for living organisms.[80]

Rivers draining agricultural lands in the Lake Erie basin are likely to be contaminated by nitrogen and phosphorus fertilizers and by pesticides. The herbicides atrazine, alachlor, metolachlor, metribuzin, and cyanazine and the oil-based organophosphate insecticides carbaryl, chlorpyrifos, carbofuran, terbufos, and fonofos all occur in locally high concentrations.[81]

Despite these grim findings, the NAWQA report does provide grounds for hope. Surface water quality in the basin has generally improved since the 1960s as some of the most egregious sources of contamination have been contained or removed. What have still to be controlled are nonpoint sources of contamination from farmlands, cities and even forested lands that were previously sprayed with DDT or other poisons, as well as industrial point sources. During a visit to the upper portion of the Cuyahoga River in 2002, I watched beaver, kingfishers, Canada geese, and a great blue heron along the portion of the river that is in the Cuyahoga Valley National Recreation Area. But a few miles downstream, I visited areas that remain a biological wasteland.[82]

Most disheartening about the Great Lakes situation is that existing regulations are not enforced. In 2001, the Bush administration proposed to cut federal environmental enforcement operations and shift resources to the states. Opposition to this proposal used Ohio as an example that states are often unable or unwilling to enforce existing laws. The Republican administrations of Governors George Voinovich and Bob Taft promoted an industry-friendly policy of voluntary environmental compliance. As a result, 72 percent of Ohio plants and refineries surveyed by the EPA during 1999–2001 had violations of the Clean Water Act. More than a third of major factories operated with expired permits required under the Clean Water Act. And when cleaning up toxic-waste sites, Ohio has averaged only one site a year for the past decade. Twelve hundred sites remain.[83]

The river biota of the Lake Erie region are affected by loss of habitat, competition from introduced species, and water withdrawals, as well as by toxic contaminants. Both streambed sediments and fish tissues contain PCBs and organochlorine pesticides, and these substances are

Top: The upper Cuyahoga River in the Cuyahoga Valley National Recreation Area, July 2002. Although impacted by agricultural and urban runoff, this portion of the river is relatively clean, in part because the floodplain and riverside corridor of vegetation perform some filtering of incoming surface waters and groundwaters. Bottom: The lower Cuyahoga River in Cleveland near the Interstate 90 bridge, July 2002. In contrast to upstream reaches, this portion of the river remains heavily polluted from urban waste. The river is completely channelized and has no floodplain or riverside forest.

passed on from the fish to birds such as gulls and eagles. As of 1998, the basin had five extinct species of fish, with an additional twenty-two endangered and nine threatened species.[84]

This region, which is an industrial hub, was once a biological treasure. Several waterfowl flyways converge on the Lake Erie–Lake St. Clair region. Hundreds of thousands of ducks, geese, and swans used the numerous wetlands of the region for resting, feeding, and breeding, and they in turn supported human hunters. From at least 100 B.C. until the arrival of the French in the late 1600s, Native Americans such as the Chippewa and Ottawa hunted and fished around the lakes. The Native Americans harvested wild rice from the wetlands and sweet grass for baskets. From the 1850s onward, the draining and loss of these wetlands accelerated with the spread first of agriculture and, after 1900, of industry and urbanization.[85]

Loss of wetland and stream habitat, increased fishing and hunting pressures, and increasing contamination levels caused substantial declines in many native species. Commercial catches of lake sturgeon in the St. Clair–Detroit River system declined from 110,000 pounds in 1870–79 to 13,000 pounds in 1960–69. The overwintering population of canvasback ducks in the eastern United States, including the Great Lakes region, dropped from more than 400,000 birds in the early 1950s to less than 148,000 by 1960 and thereafter varied between 130,000 and 280,000. It is difficult to fully realize the biological impoverishment we have caused.[86]

Hellbenders, Allegheny Alligators, and Devil Dogs

Entire species have also vanished from the Great Lakes region. One of these may be the hellbender salamander, *Cryptobranchus alleganiensis*. These salamanders are still present in the southern portion of Ohio, although they are so rare that they have been recommended for endangered status. Hellbenders may have also been present historically in the rivers draining to Lake Erie, but they are no longer found there.[87]

The scientific name for hellbenders reflects their internal gills (*Cryptobranchus* is from the Greek *kryptos* for secret or hidden and *branchia* for gills) and their presence in streams of the Allegheny Mountains. As a salamander, the hellbender is an amphibian, a word that comes from the Greek for "double life." When amphibians evolved nearly four

A hellbender salamander. (Courtesy of the Missouri Department of Conservation.)

hundred million years ago, they were the first animals to spend a portion of their lives on land. Modern amphibians continue to divide their lives among eggs laid in water, aquatic larvae with gills, and air-breathing adults capable of living out of the water. Hellbenders are unusual salamanders in that they remain completely aquatic.

To a human, the life of a hellbender might not seem very exciting. Turtles, water snakes, larger fish, and humans are the obvious dangers to be avoided. But mostly, the solitary salamander spends its days resting under submerged rocks and logs in a river, letting the world flow by. The world must flow, for the hellbender needs clear, fast-moving water to breathe. Unlike most amphibians, the capillaries in the hellbender's skin penetrate through to the surface cell layer, facilitating the exchange of oxygen and carbon dioxide between the water and the hellbender's blood. The hellbender has wavy folds of skin along its sides, and these have a particularly dense and extensive capillary network. The salamander breathes by swaying its body from side to side and rippling its skin folds to mix the water around its body, ensuring that oxygenated water is always present for breathing through the skin. This form of breathing has served the hellbender well for hundreds of millions of years, but now it makes the animal especially vulnerable, for contaminants bound to water molecules have the potential to pass directly into the salamander's body.[88]

At night the hellbender emerges to forage for food. It does not need to see well, and its small eyes are not very effective at forming images although they can detect light. The lack of sharp vision is compensated by light-sensitive skin, with the tail most sensitive of all. Large by salamander standards, the hellbender grows as long as twenty-nine inches, although it is more likely to be half of that length. The hellbender has a broad, flattened head and body, and its brown color blends well with the rocks on a streambed. The salamander is not exactly a fearsome predator as it walks along the stream bottom on its sturdy legs, occasionally swimming. A rough pad on the end of each toe helps provide traction on the slippery rocks. The hellbender sucks in its prey, depressing one side of its lower jaw and drawing in a jet of water and food. Crayfish are the main course, but the salamander also eats insects, small fish, and worms.

Like other amphibians, the hellbender has no internal temperature regulation and depends on outside heat sources to warm its body. When the temperature drops, the animal goes dormant. This strategy has the advantage of lower energy requirements. An amphibian may require less than 5 percent of the food calories used by an equal-sized mammal over a given time because the mammal has to fuel its internal heat engine. But before it goes dormant for the winter, a hellbender puts on a burst of daytime activity during the autumn breeding season. Each male excavates a nest site beneath flat rocks or other debris and then aggressively defends the site from competitors and predators. He attracts a female, and she lays spherical eggs a fraction of an inch in diameter in paired strings that form an egg mass of up to 450 eggs. The male may attract more than one female to the site, amassing more than 1,000 eggs in the nest. These eggs he fertilizes, undulating his lower body while floating over the egg mass in order to disperse sperm throughout the nest cavity. The male then guards the eggs for two to three months until they hatch into inch-long larvae. The young salamanders grow slowly, adding about an inch each year. They reach sexual maturity at five to six years of age and may live up to twenty-nine years.

Thus the quiet life of a hellbender. What is perhaps most impressive is the persistence of the species. Hellbenders belong to the suborder Cryptobranchoidea, the most primitive of the living salamanders. They are elders among salamanders in all senses of the word—large-bodied, long-lived creatures of ancient lineage. To persist this long, they have adjusted to many changes. Undoubtedly, the greatest was that crucial pioneering step of leaving the water to occupy land. Since that first giant step, salamanders have continually adapted as mountain ranges were uplifted and eroded and glaciers advanced and retreated, rearranging whole river systems.

The northern end of the hellbender's range in the eastern and midwestern United States was covered by the continental ice sheet between twenty and ten thousand years ago. The Great Lakes region was free of glacial ice by ninety-five hundred years ago, but several thousand more years passed before modern communities of reptiles and amphibians migrated into the area from elsewhere. During this period the Earth's crust slowly rebounded upward after being depressed beneath the weight of the ice, and modern drainage patterns became established.

As the glaciers retreated, rivers disrupted by ice were recolonized by hellbenders from northward-flowing rivers. Scientists using mitochondrial DNA as a marker inferred that hellbenders from the Ohio River drainage basin invaded the streams of the Ozarks. Hellbenders can disperse only between upland areas when lowland rivers are relatively clear because the salamanders cannot travel long distances over dry land, and they cannot tolerate turbid, silty waters. Imagine the hellbenders in a world of rivers newly freed of ice, steadily migrating outward from their core range, crawling up riffles, swimming down pools, crossing a short distance overland when necessary—intrepid pioneers.[89]

Today, the pioneers are fighting for a toehold. Endangered in Ohio, they have not been collected from their Hamilton County core range of the Ohio, Great Miami, and Whitewater Rivers since 1961. Their numbers have declined markedly in Indiana since 1948. In western Virginia, chemical pollution in the Tennessee River drainage basin poses a serious threat. Throughout their range, hellbenders are menaced by agricultural runoff, acidic runoff from large-scale mining, indiscriminate collecting for the pet trade and scientific research, and, most serious of all, the impoundment of rivers and streams for recreational lakes and hydroelectric facilities.[90]

Folklore describes hellbenders smearing the lines of fishers with slime in an attempt to drive them out of an area. Other stories claim that the hellbender will chase off game fish or inflict a poisonous bite if disturbed. The salamander has been called the Allegheny Alligator and Devil Dog. The hellbender, however, is not poisonous, and if it encounters game fish, it is likely to be eaten. The hellbender is harmless to anything larger than a crayfish, but like other salamanders it touches off an atavistic aversion in some humans. Far from sliming a fisher's line, the hellbender's life-giving mucus is covered by the fine sediments released into streams by human activities, and the hellbender smothers. This animal that evolved nearly four hundred million years ago, and withstood the rearrangement of continents, the rising of mountains, and the advances and retreats of glacial ice, is now vanishing beneath the loads of sediment and chemical pollution, and the locks and dams, with which we assault our rivers. The hellbender's capillary sensitivity to the water in which it lives parallels a river's sensitivity to the surrounding landscape.

The river exploitation begun by the pioneers and amplified through commercial activities was enthusiastically supported by various levels of government. From county and state agencies to federal organizations such as the Bureau of Reclamation, the Army Corps of Engineers, and the Soil Conservation Service, the government set out to enforce its jurisdiction on rivers. Rivers were dammed, diverted, straightened, channelized, confined within levees, stabilized by bank-protection or grade-control structures, bridged, and calmed through energy-dissipation structures. The government brought massive resources and expertise to the task of altering rivers to meet human needs.

Chapter 5

Institutional Conquest

Bureaucratic Impacts

A government must eventually respond to the demands of its citizens if it is to remain in power. In the United States, these demands have included some aspect of river control for two hundred years. The construction of local, discontinuous levees along the lower Mississippi River by landowners during the early 1700s, for example, led to so many conflicts of interest among landowners that Secretary of War John C. Calhoun recommended that the Army Corps of Engineers take control in 1819. In 1824 Congress passed the Rivers and Harbors Bill authorizing the removal of sandbars and wood along the Mississippi and Ohio Rivers. This was followed by the 1849 and 1850 Swamp Acts, which granted to the states all unsold swamplands and provided that funds from the sale of those lands be used for flood protection. The irony of these acts is painful today, as we increasingly realize the importance of swamplands in mitigating floods. The massive 1858 flood on the Mississippi River resulted in a flood-control policy focused on levees. In the last decades of the nineteenth century, drainage districts were established across the United States, and thousands of miles of trenches were gouged to dry up wetlands. The U.S. Congress directed the Corps of Engineers to dredge a channel six feet deep from the mouth of the Mississippi to Minneapolis in the 1900s.[1]

John Wesley Powell began arguing for governmental watershed surveys and reclamation projects in the western United States in the 1870s. His work resulted in the Newlands Act of 1902, which established the Bureau of Reclamation with the intent of "reclaiming" arid regions for agriculture by regulating and storing the flow of rivers.[2]

After these early actions, the intervention of the federal government in river processes did not substantially accelerate until the 1927 flood on the Mississippi River and the New Deal programs of the 1930s. The flood of spring 1927 displaced nearly 1 million people at a time when the national population was only 120 million, killed an estimated 1,000 people, and caused $1 billion in total losses. The resulting 1928 Flood Control Act, with $300 million in federal funds, made flood control in the lower Mississippi River valley a federal responsibility. Within a few years, the Roosevelt administration enacted a series of broad programs designed to lift the country from its economic depression. The idea became widely accepted that the federal government was responsible for engineering natural environments in order to mitigate natural disasters and to increase agricultural and industrial productivity. Both the Dust Bowl droughts and periodic severe floods created national socioeconomic disasters. Americans decided that the government must prevent future disasters with massive engineering projects to tame rivers and store their floodwaters for dry years. The number of civil engineers in the United States rose steadily during the 1920s and 1930s. This increase in engineering, combined with ideas of scientific management and the perceived inefficiencies of natural systems, led to much more intensive and intrusive management strategies for rivers. Simultaneously, the government altered watersheds in its drive to raise the national standard of living by improving the road network and providing such utilities as electricity, telephones, and sewer and water lines for everyone.[3]

Three federal agencies were primarily charged with the tasks of altering river processes. The Bureau of Reclamation, created in 1902, was given the task of reclaiming arid lands in the western United States. The bureau is now most noted for the construction of huge reservoirs on western rivers. The Army Corps of Engineers was charged with the improvement of waterways for navigation and flood control. The corps is now most noted for channelization, levee construction, and dam building along eastern and midwestern rivers such as the Mississippi River. The Department of Agriculture's Soil Conservation Service, formed in 1935, was charged with flood control and land reclamation along smaller, headwater channels. The service is most noted for constructing sediment-detention and erosion-control structures along these smaller streams.[4]

The legislation enabling these agencies to modify river processes largely stems from the 1936 Flood Control Act. This act, and the successive flood control acts of 1948, 1954, 1966, and 1970, dramatically changed the American landscape. The 1936 act authorized the Corps of Engineers, in cooperation with local governments, to plan and participate in flood control improvements of major drainages on navigable waters and their tributaries. In practice, this meant that meandering, irregular rivers that periodically flooded adjoining valley bottoms were to be converted to straight canals, bordered by levees, to efficiently convey both boat traffic and floodwaters downstream to the ocean.[5]

In terms of miles of river straightened and acre-feet of water stored in reservoirs, the federal agencies were spectacularly successful in their river engineering operations. From 1820 to 1970, more than 200,000 miles of waterways were modified to reduce flooding, drain land for agricultural use, and provide for the waterborne transport of goods. About 130 million acres of wetlands were drained. Between 1940 and 1970, the Corps of Engineers and the Soil Conservation Service alone modified 34,240 miles of waterways in 1,630 projects. In the 75,000 square miles of the upper Mississippi River basin, 8,000 miles of levees were built, and 65 percent of the original wetlands were drained. Twenty-nine locks and dams were built along the upper Mississippi River from St. Louis to Minneapolis by 1950, and a nine-foot navigation channel was dredged along the Missouri River from St. Louis to Sioux City, Iowa. Single projects were massive in scope. The Tennessee-Tombigbee Canal, authorized in 1946 and completed in the 1980s, focused on a 232-mile-long canal that the corps built to link the two rivers. In the process, the corps dredged and excavated more than twice the material removed to build the Panama Canal. The Tennessee-Tombigbee Canal created a navigational shortcut for an existing barge route from Tennessee to the Gulf of Mexico. At a cost of approximately $4 billion and 100,000 acres of forests and agricultural lands, the canal offers cheaper shipping rates and subsidizes the oil and coal companies that constitute 70 percent of the canal's traffic.[6]

By the end of the twentieth century, more than seventy-five thousand large dams were built in the continental United States. Large dams in this context are greater than six feet tall and have more than fifty acre-feet of storage. In some regions of the country, such as the Northeast, an estimated ten small dams were built for every large dam. In the early 1930s alone, the Civilian Conservation Corps installed more than thirty-one thousand in-stream structures, many of which were small dams. Existing large dams were capable of storing a volume of water almost equaling the average runoff during an entire year throughout the country. In the Great Plains, the Rocky Mountains, and the arid Southwest, large dams store nearly four times the mean annual runoff. No one has quantified the storage of the small dams.[7]

In 1960, 3,123 large dams were completed, the greatest number in one year. The major river systems of the country were altered from free-flowing networks to stoppered ponds. Between 1933 and 1975, twenty-eight large dams were built on the rivers draining the 233,000 square miles of the Columbia River basin. Rivers in the 227,000 square miles of the Colorado River basin were rearranged by nineteen major dams and eight water-export projects. The Tennessee Valley Authority built thirty-seven large dams for hydroelectric power and flood control across the 28,000 square miles within its jurisdiction.[8]

Federal modification of the nation's rivers continued rapidly from the 1930s onward. But by the late 1960s, a federal task force on federal flood control policy noted that, despite a federal investment of more than $7 billion (in 1967 dollars) in flood control projects since 1936, the nation's flood damage bill averaged roughly $1 billion annually. As recognition grew that the structural approach was inadequate for controlling flooding, attention to alternatives such as land-use zoning and flood insurance increased. Congress established the National Flood Insurance Program in 1968. The 1970 National Environmental Protection Act provided a framework for land-use zoning. In 1973, a Congressional Committee on Government Operations produced the report *Stream Channelization: What Federally Financed Draglines and Bulldozers Do to Our Nation's Streams*. The report described "a traumatic assault against free-flowing water bodies whose natural resources are often irreplaceable." And in 1974 the Streambank Erosion Control Evaluation and Demonstration Act authorized a five-year program by the Corps of Engineers to define the magnitude of the problem of streambank erosion, identify the causes of erosion, and evaluate the most promising methods of bank protection. After four years of study, the corps estimated that total damages associated with streambank erosion ran to $270 million per year and remedial costs using conventional methods reached $870 million per year. The combination of escalating costs from floods and other resource damage, and the growing environmental movement, was catching up with federally financed river engineering.[9]

Reconsidering Dams

A large body of technical and popular literature describes the negative effects of dams on river ecosystems and the changes in public perception of dams during the twentieth century. Flow is a major determinant of river form and habitat, which in turn are major determinants of ecological communities. In-channel and riverside species have evolved life history strategies primarily in response to natural flow regimes, and alteration of these regimes stresses or eliminates species and communities. Dams and flow regulation alter the upstream-downstream and channel-floodplain connectivity that are essential to many riverine species and may favor the invasion of exotic and introduced species.[10]

The size and purpose of a dam partly control how a dam affects a river. Dams built on rivers in the eastern United States during the eighteenth and nineteenth centuries included very low structures designed to catch sawn logs being transported downstream, and tall structures designed to produce a drop used to power a mill. Neither type was designed to impound a large volume of water. Water-storage dams built in dry regions were sited to maximize water storage, and their operation tends to reduce spring peak flows and increase late summer base flows. Dams built for hydroelectric power generation are operated in response to daily, weekly, and seasonal power demands, creating highly irregular flows downstream. Differences in dam design and operation produce different impacts on water flow, sediment and nutrient movement, and fish passage.

The Grand Canyon of the Colorado River provides an example of the impacts of dams. When settlers of European American descent first reached the Grand Canyon during the 1870s and 1880s, they mined various ores along the river and planned to build a railroad beside the channel. In 1919, during the commercial era, recognition of the canyon's scenic value led to its designation as a national park. There was little public opposition when Glen Canyon Dam was built just upstream from the national park boundary in 1963. Some environmentalists were outraged at the flooding of scenic Glen Canyon, but most Americans approved of the dam construction as providing water storage, hydroelectric

View of Glen Canyon Dam on the Colorado River from downstream. Besides visually dominating the river corridor, the dam's regulation of the flow of water, sediment, and nutrients affects the river ecosystem for tens of miles downstream.

power generation, and increased recreational potential. Lake Powell, impounded behind the dam, became a popular recreational site for fishing, water skiing, and houseboat outings. Native fish, many of which are now endangered, were poisoned with rotenone, and the lake was stocked with nonnative game fish.[11]

Glen Canyon Dam was one of the last huge dams built by the U.S. government that faced minimal public opposition. During the 1960s and 1970s, public opinion gradually shifted toward environmental protection and against the regulation of rivers by dams. As the average educational and socioeconomic level of American society increased, affluent urban and suburban dwellers with leisure time put more value on increasingly scarce wilderness. Growing popular support for outdoor recreational opportunities in natural environments, and for environmental quality, translated into increasing support for governmental regulation of resource use. Citizen groups such as the Sierra Club and the National Audubon Society developed sufficient membership to push congressional representatives to support legislation including the Wilderness Act (1964), the Wild and Scenic Rivers Act (1968), the

Endangered Species Act (1973), and the Clean Water Act (1977). These acts made it increasingly difficult for the federal government to approve newly proposed river regulation projects. When another dam was proposed within Grand Canyon National Monument in 1966, the public outcry effectively stopped the project. Dam promoters claimed that the dam's reservoir would make the rugged Grand Canyon more accessible for tourism. The Sierra Club responded with a full-page advertisement in the *New York Times* asking whether we should flood the Sistine Chapel to make Michelangelo's ceiling paintings more accessible.[12]

Glen Canyon Dam makes little sense from a hydrologic standpoint. Evaporative losses from the long reservoir behind the dam, and seepage losses into the porous sandstone surrounding the reservoir, are estimated at 882,000 acre-feet per year. This is more than 6 percent of the average annual water yield of the Colorado River system, and almost three times Nevada's annual 300,000 acre-feet entitlement from the Colorado River. Along with other consumptive uses of water in the desert Colorado River basin, these losses help to prevent the Colorado from reaching its historical delta in the Gulf of California.[13]

Since the completion of the Glen Canyon Dam in 1963, numerous scientific studies have explored the dam's impact on the Colorado River ecosystem. The Colorado River historically had warm, sediment-laden waters. The river flowed high during the snowmelt of late spring and early summer and then subsided to much lower flows for the remainder of the year. Native aquatic and riverside plants and animals evolved in association with these conditions. As with many dams, Glen Canyon Dam creates a giant settling pond for sediment upstream, and the waters released from the base of the dam are clear and cold. The annual flood peak is reduced by storage in the reservoir, and base flow during the remainder of the year is increased above historical levels. The flow in the river also fluctuates dramatically on twenty-four-hour and weekly cycles as demand for hydroelectric power fluctuates.[14]

The changes in water temperature, chemistry, sediment load, and flow favor introduced game fish such as trout at the expense of native species such as the razorback sucker and Colorado pikeminnow. The native species are now mostly listed as threatened or endangered. The reduction in downstream sediment transport led to the erosion of beaches used for camping by the tens of thousands of people who annually float the Colorado River through Grand Canyon and the erosion

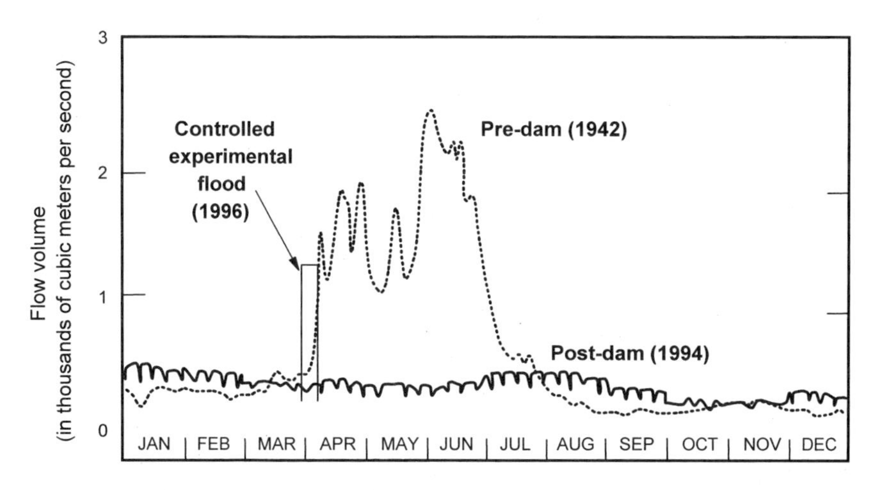

Comparison of the experimental flood of 1996 with a typical annual flow pattern (1942) before construction of the Glen Canyon Dam and a typical annual flow pattern (1994) after construction of the dam. (After A. D. Konieczki, J. B. Graf, and M. C. Carpenter, 1997, Streamflow and sediment data collected to determine the effects of a controlled flood in March and April 1996 on the Colorado River between Lees Ferry and Diamond Creek, Arizona, *U.S. Geological Survey Open-File Report 97-224*, Figure 2.)

of river terraces containing archeological sites. Flash floods and debris flows periodically coming down the tributaries to the Colorado River create tributary debris fans that constrict the Colorado River and cause rapids. Large floods on the Colorado are capable of eroding these fans and reducing the rapids, but as Glen Canyon Dam reduces flood flows, this effect is lost and the rapids grow more dangerous to boaters.[15]

Recognition of these downstream changes prompted the idea that occasional releases of flood flows from the dam might restore some of the downstream river ecosystem. A large snowpack and rapid snowmelt in 1983 created an inadvertent flood release that severely damaged the dam spillways. An experimental flood release during March 1996 was of smaller magnitude than the 1983 release but was accompanied by numerous scientific studies. The 1996 flood was widely publicized as returning the Colorado River to its historical state, although the flood was small and of short duration when compared with natural, pre-dam floods. Nonetheless, the deliberate release of impounded waters with the intent of restoring river ecosystem functions marked a turning point in river regulation.[16]

This change in attitude toward rivers was at least in part driven by scientific understanding of river ecosystems, and by dissemination of that understanding to the public. Studies of the physical, chemical, and biological processes operating along rivers gradually led scientists to recognize the complexity and interdependence among components in a river ecosystem. River food webs, for example, begin with primary production from photosynthetic plants, or with aquatic insects processing organic litter dropped into the stream from overhanging tree canopies. Rates of primary production depend on water temperature, turbidity, chemistry, and flow. Primary production also depends on the plant community growing beside the channel, as well as exchange between river water and groundwater. Valley geology, form, and climate influence primary production as they affect riverside plant communities, water and sediment yield to the river, groundwater exchange, and so forth. Downstream food webs also depend on inputs from upstream. Try to isolate any component of a river ecosystem, and you find it inextricably linked to many other components.

At the beginning of the twenty-first century, the era of federal sponsorship for building large dams in the United States has ended. There is widespread public recognition that dams constructed during the heyday of dam building from the 1930s through the 1970s have adversely affected American rivers, and societal support is growing for modifying or removing dams in order to restore river ecosystem functions. The Web site for the nonprofit organization American Rivers lists 485 small dams that had been removed as of 2003, such as the Edwards Dam on the Kennebec River in Maine.

The widespread construction of levees also came under serious review after floods on the Mississippi River in 1993. These floods produced an estimated $18 billion in damage and killed fifty-two people. Thousands of miles of levees were built along rivers in the United States to contain floodwaters and improve channel conveyance, but levees also isolate a river from its floodplain, with negative ecological impacts. The levees exacerbate downstream flooding by preventing floodwaters from moving slowly downstream across floodplains, and they also create downstream sedimentation problems as confined, energetic floodwaters erode the riverbed and banks. After the 1993 flood, greater public and governmental consideration was given to alternatives, such as levees set back from the channel along a floodplain, or reduction of floodplain land uses that require flood protection. However, the national debate did not lead to rapid changes in policy or the widespread implementation of alternative forms of flood control. Attitudes changed slowly, if at all, within the Corps of Engineers, as reflected in the comments of a spokesperson during 2001: "Floods aren't caused by levees, the loss of wetlands, navigation structures in river or floodplain development. . . . What causes floods is a lot of rainfall." Certainly a lot of rainfall causes floods, but the spokesperson's comments ignore the role that levees, loss of wetlands, navigation structures, and floodplain development play in exacerbating floods and flood damages. Adjusted for inflation, the cost of flood damages continues to rise dramatically in the United States because of continuing human encroachment on rivers and floodplains.[17]

More "Efficient" Rivers: Channelization

The third major form of governmental alteration of river systems, channelization, has largely escaped the public scrutiny recently given to dams and levees. Channelization is the widening, deepening, clearing, and/or straightening of river channels. The bed and banks of the channelized river are sometimes covered with concrete or with rock riprap, a layer of rocks larger than those that can be moved by the flow of the river. Such activities are undertaken for several purposes. Channelization drains wetlands by speeding the passage of water through the wetlands and lowering the groundwater table. Channelization reduces flooding of adjacent lands by increasing the river's capacity to transport flood flows, or enhances navigation by increasing the natural depth of larger rivers. Channelization also controls erosion by substituting artificial canals for gullies or other eroding natural channels. Individuals or local communities have undertaken channelization for two centuries in the United States, but channelization became much more extensive under the supervision of the federal government during the 1940s. Of the more than thirty-four thousand miles of waterways channelized by the Army Corps of Engineers and the Soil Conservation Service after 1940, about 8 percent were in only fifteen states. The southern states of North Carolina, Tennessee, Georgia, Louisiana, Mississippi, and Alabama and the midwestern states of Illinois, Indiana, North Dakota, Ohio, and Kansas, along with California and Florida, bore the brunt of channel-alteration work. Five states had about half of the total number of projects.[18]

A 1997 aerial photograph of a channelized river seventy miles northeast of Memphis, Tennessee. The original river channel is the sinuous lighter-colored line, and the channelized stream is the straight line below it.

It became apparent within thirty years that channelization had some unanticipated consequences. A 1973 congressional report noted that most of the open ditches constructed before 1940 to drain wetlands were poorly engineered, poorly maintained, and poorly designed in relation to their larger watersheds. Consequently, the local entities constructing ditches solved the flood problem by dumping it downstream. The federally financed channelization projects of the 1940s to 1970s, which used large bulldozers and draglines, were not much better. The 1973 report on these more recent projects is damning. The report emphasizes that "inadequate consideration was being given to the adverse environmental effects of channelization. Indeed, there is considerable evidence that little was known about these effects and, even more disturbing, little was done to ascertain them."[19]

Scientists have demonstrated the adverse environmental effects of channelization to be many and various. In addition to loss of upland soil, adverse lowland effects include impacts to wetlands, riverside vegetation, river form and flood flow, and aquatic organisms.

The Drying Game

Drainage of wetlands lowers the local groundwater table, changing the water cycle and availability of nutrients for wetland plants. This eliminates or reduces both the number and diversity of plant and animal species living in and using the wetlands. As the water table declines, the water-holding capacity and the capacity for groundwater recharge also decline. The Des Plaines River of Illinois, mentioned in the previous chapter in connection with water pollution, provides an example of wetlands loss. This river has been ill-used. Besides being polluted with nonpoint-source contaminants from both urban and agricultural activities, portions of the streambed were channelized, the banks cut steep, and the floodplain leveed. As a result, homes downstream are subject to more frequent flooding. Wetlands along the river were drained with tile fields and mined for gravel. Studies of fish communities in the Des Plaines River indicate a decrease in species diversity since the early period of European American settlement. Ninety percent of the fish biomass in the altered segments of the river is now carp. The original wetlands and their associated plant and animal species are gone.[20]

Although channelization initially drains wetlands, erosion of upstream channel segments that are adjusting to channelization can produce large quantities of sediment that fill downstream channel segments. This sedimentation can so decrease channel capacity that overbank flooding increases relative to conditions before channelization. Extreme loss of channel capacity can change seasonally inundated floodplains into huge marshes.

Simplify, Simplify

Periodic flooding and lateral channel movement are natural disturbances that create spatial variability in bottomland forests. On the southeastern coastal plain, for example, the dynamics of the channel and floodplain create a complex mosaic of vegetation. Black willow, water tupelo, and bald cypress grow in the abandoned meanders that hold water most of the year. New surfaces are created as the channel moves sideways, depositing point bars and filling old, abandoned meanders. These new surfaces are rapidly colonized by flood-tolerant, opportu-

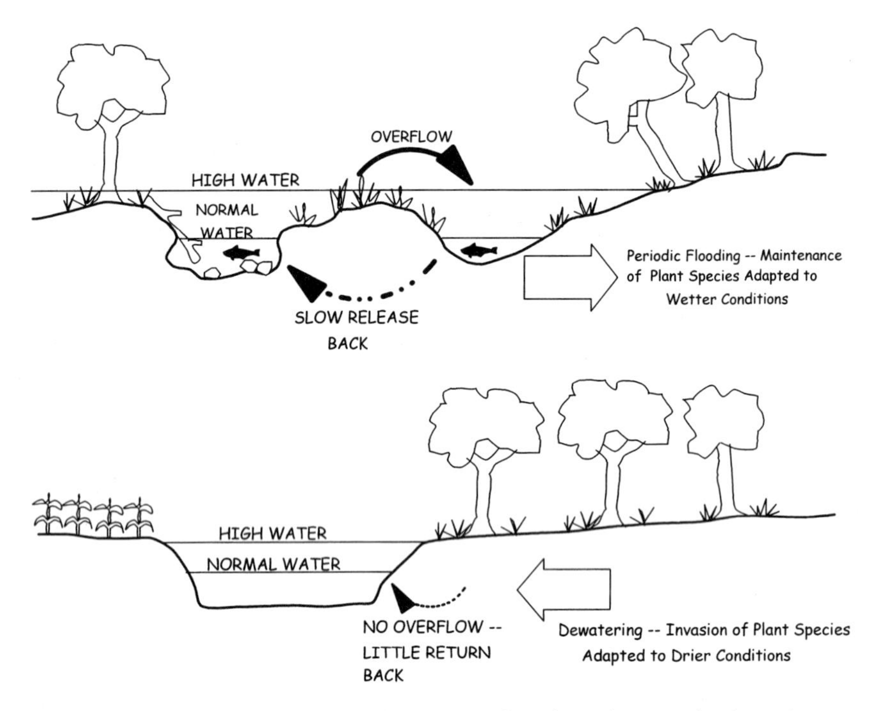

Comparison of natural and channelized streams. Before channelization, the channel is bordered by wetlands that experience seasonal flooding. These wetlands store floodwaters and gradually return them to the channel. The periodic flooding supports a variety of plant species adapted to wetter conditions. After channelization, floodwaters are contained within the main channel. The wetlands dry out and are invaded by plant species adapted to drier conditions. Land use can also change along the channel banks. (After P. W. Simpson, J. R. Newman, M. A. Keirn, R. M. Matter, and P. A. Guthrie, 1982, *Manual of stream channelization impacts on fish and wildlife,* FWS/OBS-82/24, Figures 4-5, 5-10, and 5-11.)

nistic species such as black willow, cottonwood, and silver maple. The streams flood most years in the winter and spring, repeatedly submerging portions of the floodplain for a few days to several weeks. Sediment deposited during flooding raises the elevation of new point-bar surfaces, making them less susceptible to inundation and more suitable for species such as green ash, sugarberry, water hickory, and water locust. When channel obstructions force water into the adjacent bottomlands, shallow swamps destroy forest stands, creating new sites for the establishment of early successional species. A bird flying over the southeastern bottomlands would thus see small patches of shallow marsh, tupelo and cypress swamps, willow and maple stands, and ash and hickory woodlands all mixed together. But stabilizing a river reduces variability through time and across the bottomland. Stabilized channels preclude the development of shallow swamps, point bars, and abandoned meanders. Eventually, the bottomland forest shifts toward a homogeneous community of later successional species less tolerant of flooding, which occupied the outer floodplain before channelization.[21]

Fast and Loose

Direct cutting of riverside trees during channelization eliminates shading and the input of organic matter such as leaves and twigs to streams. With the trees and their binding roots gone, the streambanks are more susceptible to erosion. Flow velocity is not as effectively reduced along the now-smooth streambanks, and sediment is less likely to be deposited in natural levees. Cutting of bottomland hardwoods eliminates vital habitat for many animals and can increase nutrient and sediment concentrations in adjacent stream channels.

Scientists have described the evolution of stream channels in lowland Mississippi after channelization using a six-stage model. In stage I, before channel modification, the bed and banks of the meandering river are stable or accumulating sediment, with minimal large failures along the banks. The river has a mature, diverse riverside community and complete vegetation cover. Stage II represents channel straightening. The straight channel has linear banks and a steep downstream slope. All woody vegetation is typically removed. This stage usually lasts less than a year before the steep slope and rapid flow cause the channel bed to erode during stage III. After one to three years of cutting down, the streambanks become unstable and begin to erode along with the streambed. Bank failure produces steeply curving banks and severe instability. The loss of riverside forests that had stabilized the banks facilitates bank erosion during this fourth stage, which may last five to fifteen years. Woody plants not directly affected by channelization are now removed by bank-slope failure. The wood introduced to the channel during this stage floats downstream to accumulate in logjams that can deflect flow and locally accelerate bed and bank erosion, or plug the channel and cause it to overflow. Eventually, sediment coming from upstream erosion begins to accumulate on the streambed during stage V. Bank failure declines. As sediment accumulates along the streambanks, riverside trees begin to reestablish. After fifty to seventy years the channel is once again stable (stage VI), with a meandering form, low banks, a lower downstream slope of the streambed, general mild sediment accumulation, point-bar development, and diverse bank vegetation. The six-stage model was developed specifically for Mississippi channels with broad, muddy floodplains. Different types of channels have slightly different stages of response and lengths of response time following channelization, but analogous stages of channel adjustment have been described for channelized streams in the Great Plains and the southwestern United States.[22]

River segments upstream and downstream from the location of channelization are also likely to be affected. Upstream segments commonly erode as a result of the steeper streambed slope. River segments below the channelization accumulate sediment because of the increased supply coming downstream from the eroding areas. Downcutting of a channel lowers the level for all of its tributaries, which erode in response, destabilizing the entire watershed. These effects may be substantial. Increases in channel width caused by channelization-related erosion range as high as 100 to 1,000 percent. Channel modifications in 1968–69 on the South Fork Forked Deer River in western Tennessee shortened channel lengths by 14 percent and increased streambed slopes by as much as 198 percent. Headward erosion at a rate of approximately one and one-half miles per year caused from five to ten feet of bed erosion along an eight-mile reach of the South Fork from 1969 to 1981. The excess sediment from erosion in turn caused seven feet of streambed accumulation downstream. This represented 60 percent of the elevation removed by the 1969 channel excavation. The Tennessee Game and Fish Commission estimated that the net economic loss from losses of wildlife, fisheries, and commercial timber associated with these channel adjustments was more than $4 million per year.[23]

During these oscillating channel adjustments, the diversity of river substrates and bedforms declines. Pools, if not gone entirely, are likely to be poorly formed and spaced relatively far apart.[24]

Channel change also damages structures such as pipelines, water intakes, and bridges. Sixty years after Missouri's Blackwater River was channelized, the cross-sectional area of the river increased up to 1,173 percent. As the channel widened and deepened, one bridge was replaced

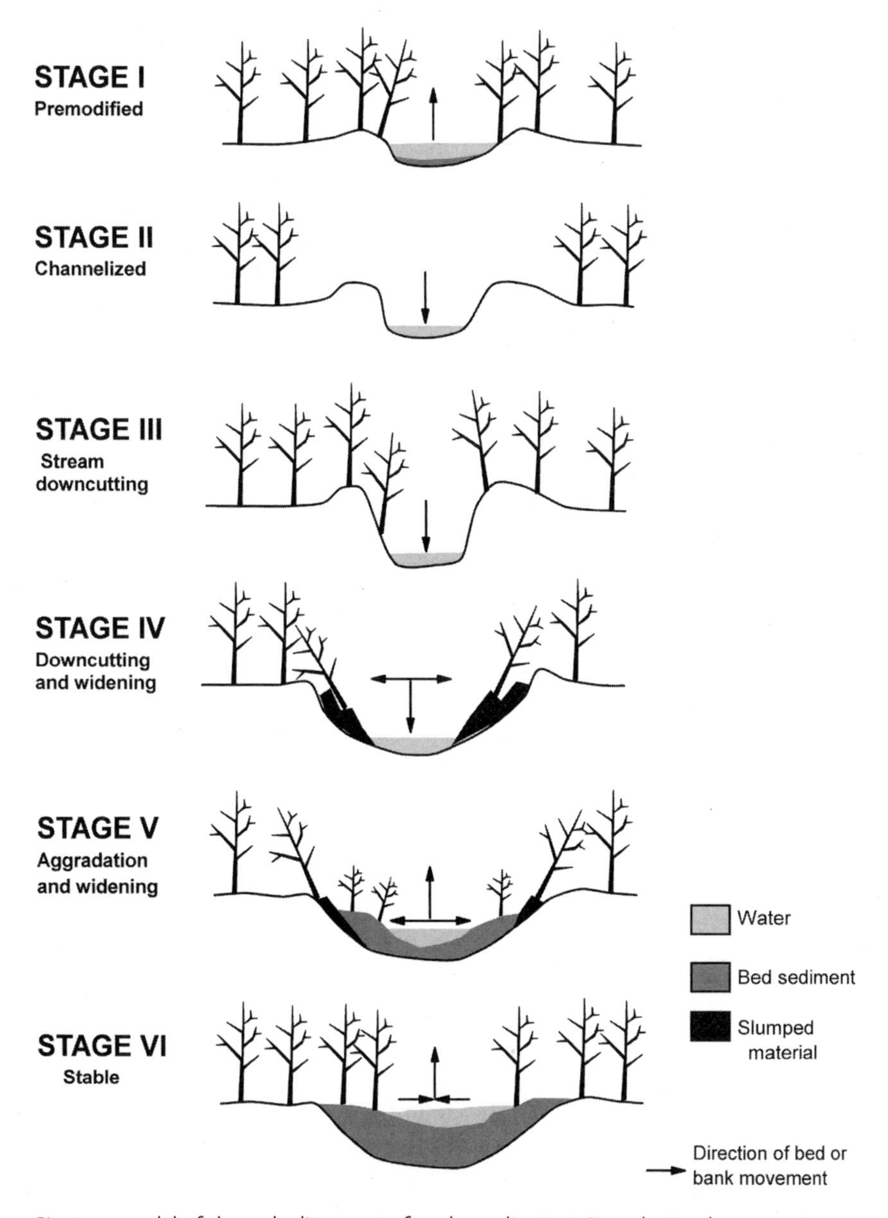

Six-stage model of channel adjustment after channelization. Sites depicted are at or just upstream from the site of channelization. (After C. R. Hupp, 1992, Riparian vegetation recovery patterns following stream channelization: A geomorphic perspective, *Ecology*, 7, 1209–26, Figure 2.)

in 1930, 1942, 1947, and again later. Severe channel erosion generally reduces farm access and income and disrupts transportation routes.[25]

Channelization tends to decrease flood duration and increase peak flood flow while the river remains channelized. A study of two adjacent streams in northwestern Mississippi found that large floods took nearly three times as long to pass along a sinuous stream as along a channelized stream. After the sinuous stream downcut in response to downstream channelization, the flood flows became similar and the benefits of downstream flood peak attenuation were lost along the sinuous stream. A study conducted nearly thirty years after channelization on the Obion River in western Tennessee indicated that the number of floods during the growing season months of May to October had increased 140 percent on the stream segments downstream from the channelization. Runoff from the channelized portion of the basin now converges at downstream locations faster than the stream channel can accommodate. Although the average duration of floods has decreased, the brief periods of inundation are still capable of destroying crops. If a channelized stream is not maintained, and loses capacity as a result of sediment from upstream, overbank flooding can become worse than it was before channelization.[26]

Carpe Stream

The removal of overhanging banks, streamside vegetation, and pools results in higher water temperatures and larger daily temperature fluctuations along channelized rivers. This in turn affects the growth rate and physiological state of fish, particularly in relation to their resistance to disease. An increase in temperature decreases dissolved oxygen, which can detrimentally affect fish. The hatching time of fish eggs, the emergence of aquatic insects, the migration of fish, and the overall productivity of the stream are also tied to water temperature. The removal of streamside vegetation reduces the input of organic matter that is an important food source for invertebrates and reduces the density of terrestrial insects that fish eat.[27]

The increased turbidity and sedimentation associated with channelization impact fish by causing inflammation of gill membranes, greater probability of infection, and reduction of the visual feeding range. Greater turbidity increases drift rates in some invertebrates and re-

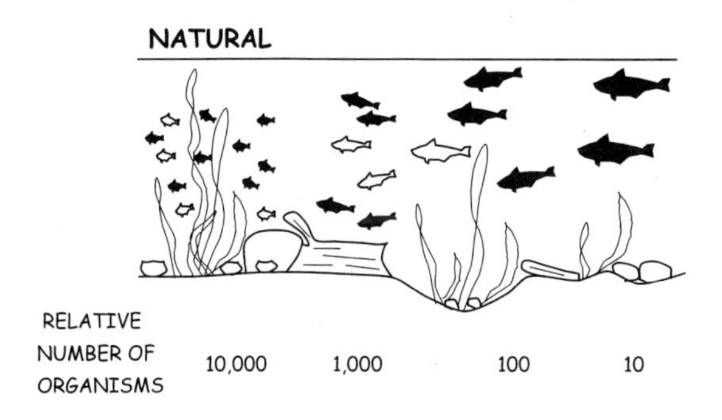

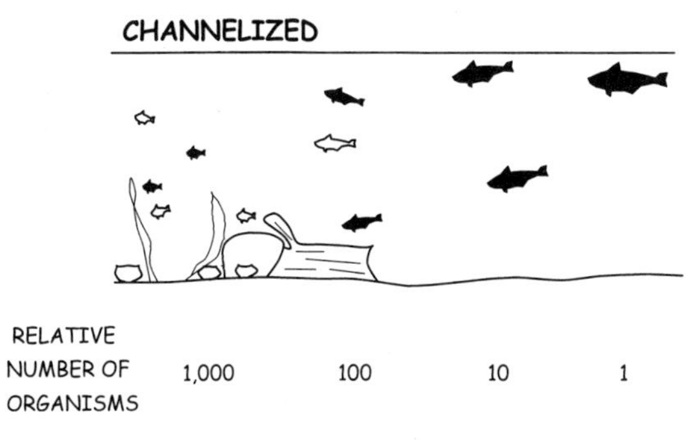

The reduced input of leaf litter, twigs, and other organic matter from channel banks reduces the basic energy supply of channelized streams and decreases the number and diversity of organisms that can be supported at each level of the food web. The relative number of organisms are shown for each level of the food web in a reach of a natural stream and a channelized stream; for each level of the food web, the number of individual organisms is reduced in a channelized stream. (After P. W. Simpson, J. R. Newman, M. A. Keirn, R. M. Matter, and P. A. Guthrie, 1982, *Manual of stream channelization impacts on fish and wildlife,* FWS/OBS-82/24, Figure 5-5.)

duces their survival, and interferes with photosynthesis in the stream. Sediment deposition destroys invertebrate habitat and invertebrate food sources of algae and attached microorganisms. Sedimentation also destroys fish spawning sites.[28]

Increased flow velocity in channelized streams affects the amount of invertebrate drift and the food supply for fish. It also erodes the streambed, destroying invertebrate habitat and reducing the survival of fish eggs. Fish species with more streamlined body types may be better able to survive in channelized streams. Other changes in the stream flow change water levels and the ability of organisms to move up- or downstream.

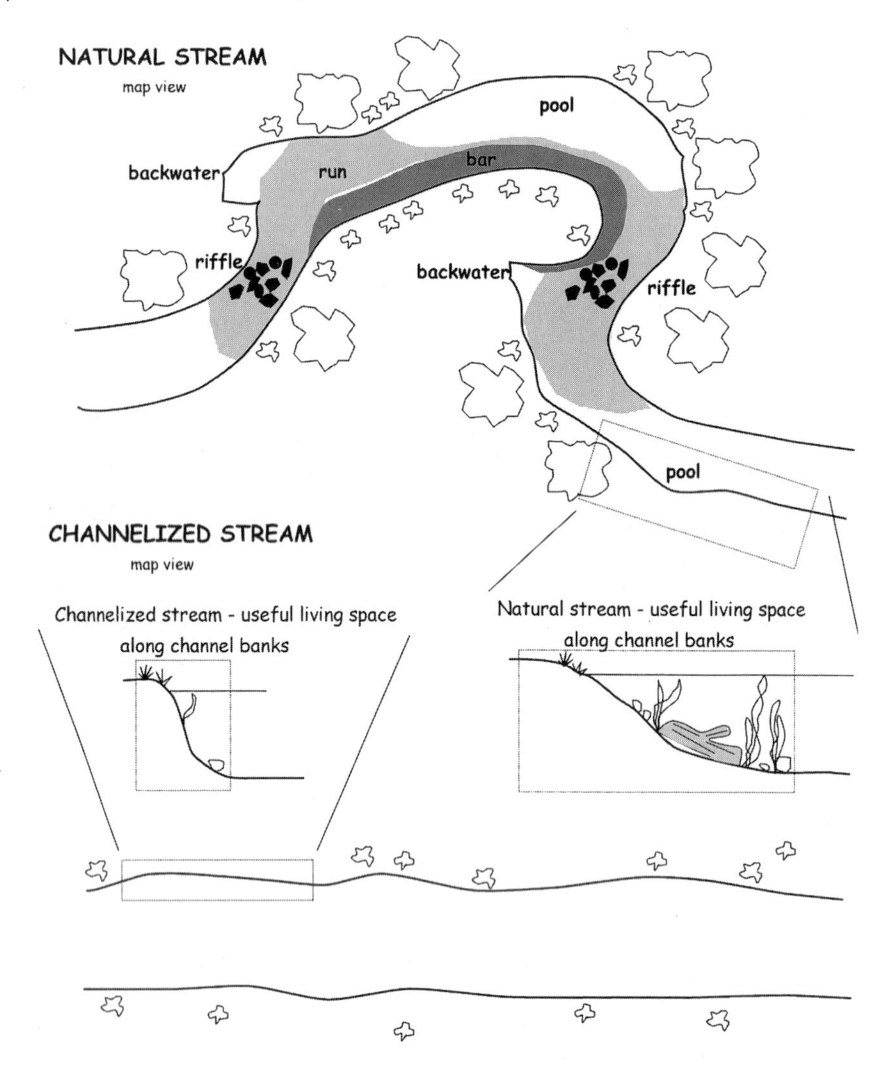

Schematic comparison of habitat diversity in natural and channelized streams. Loss of stream length and streambank complexity through channelization substantially reduces habitat availability and diversity. (After P. W. Simpson, J. R. Newman, M. A. Keirn, R. M. Matter, and P. A. Guthrie, 1982, *Manual of stream channelization impacts on fish and wildlife,* FWS/OBS-82/24, Figures 4-4 and 5-4.)

The removal of large boulders, logs, encroaching vegetation, deep pools, and undercut banks during channelization reduces shelter in the stream that provides fish with protection from predators and a place to avoid strong currents. Loss of shelter and streambed diversity decreases the diversity of invertebrates in the stream reach.[29]

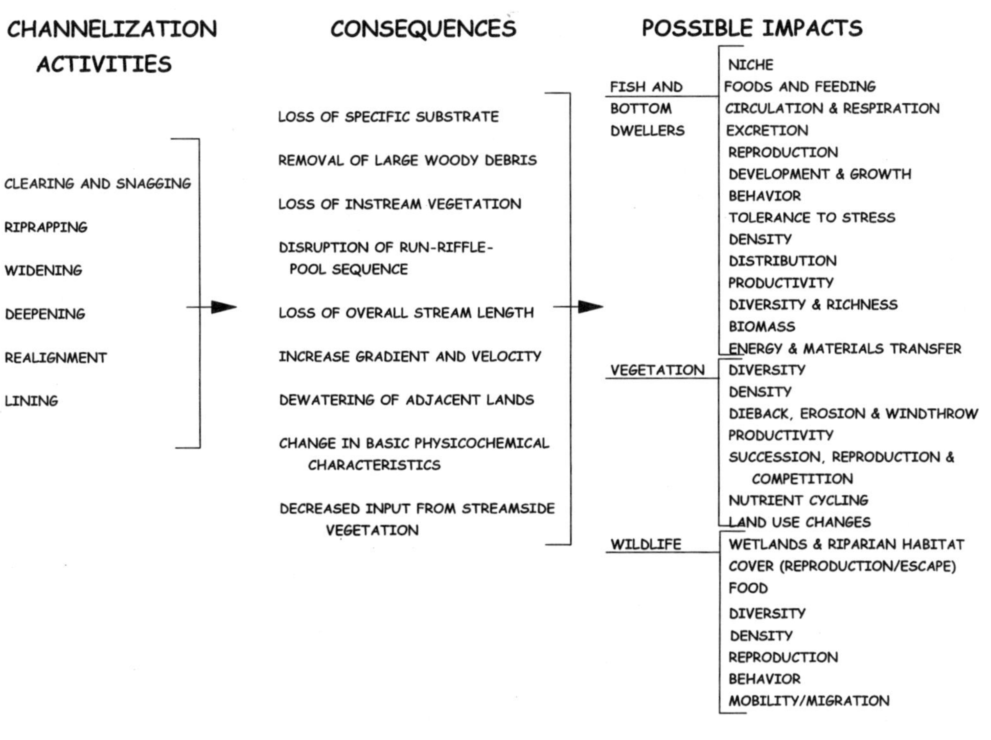

Summary of the impacts of channelization on stream and streamside communities. Categories listed under "possible impacts" are aspects of the stream ecosystem altered by channelization. "Niche," for example, indicates that details of habitat for fish and bottom-dwelling aquatic organisms are altered. (After P. W. Simpson, J. R. Newman, M. A. Keirn, R. M. Matter, and P. A. Guthrie, 1982, *Manual of stream channelization impacts on fish and wildlife,* FWS/OBS-82/24, Figure 9-1.)

During the 1860s, headwater prairie creeks in Ohio, Indiana, and Iowa rose out of broad marshy depressions, the prairie swales, which slowly contributed drainage water through perennially flowing springs. The creeks occupied narrow, highly sinuous channels less than three feet deep. Twenty years after channelization, the creek beds lie from six to more than nine feet below bank height. The straight channels are nine to thirty feet wide. Abundance, diversity, and biomass of fish and invertebrates are significantly lower in the channelized streams. Water quality is lower and habitat structure is less complex. The channelized streams carry heavier sediment and nutrient loads, have a wider range in daily flows, and are largely devoid of in-stream vegetation. Environmentally sensitive species are absent or present only in small numbers. Numerous fish species adapted to clear water, stable streambeds, and lush aquatic vegetation were replaced by tolerant forms such as carp.[30]

Nearly every study conducted on channelized rivers yields analogous results. The Little Sioux River in Iowa is a turbid, warm-water river initially channelized during 1905–1920, and again during the early 1960s. Subsequent studies showed that the channelized stream sections had higher maximum and mean daily water temperatures, greater daily temperature fluctuations in summer, a lack of suitable attachment areas for bottom-dwelling insects, and fewer fish species than unchannelized stream sections. Channelized coastal plain streams in North Carolina lost most of their pool capacity and had 215 percent less average weight of all fish per surface acre than otherwise analogous natural streams. By removing streamside vegetation along Roanoke Creek in Virginia, channelization significantly lowered the diversity and density of bird species relative to undisturbed areas.[31]

Nearly all of the 1,842 miles of major streams in the state of Missouri north of the Missouri River were channelized or are threatened with channelization or inundation by flood-control reservoirs. Comparative studies along the Chariton River in northern Missouri indicated twenty-one fish species in unchannelized portions of the river and thirteen species in channelized sections. The standing crop of fish was reduced 83 percent in the channelized sections from 304 pounds per acre to 53 pounds per acre. Similarly, fish catches were two to two-and-one-half times greater along 50 unchannelized miles of the Missouri River than along 250 unchannelized miles of the river. Sixty years after channelization, channelized reaches of the Blackwater River in Missouri had 12 pounds of large fish (longer than twelve inches) per acre compared with 403 pounds per acre in unchannelized reaches.[32]

Statistics from studies across the country clearly indicate that stream channelization does not produce the intended benefits of flood control. What channelization does clearly produce is economic and environmental devastation, often for decades after the original channel engineering.

The Downside of Efficiency

The early 1970s were a period of much negative publicity for channelization. In 1971 Assistant Secretary of the Interior Nathaniel P. Reed testified in the House of Representatives that channelization in many regions of the United States reduced local populations of fish, plants, and ducks by 80 to 99 percent and that, contrary to assertions by the Soil Conservation Service, the loss was often permanent. He stated, "Stream channel alteration under the banner of 'improvement' is undoubtedly one of the most destructive water management practices . . . the aquatic version of the dust-bowl disaster." Eugene C. Buie, assistant deputy administrator of the Soil Conservation Service, responded that "American agriculture couldn't survive without it."[33]

In 1973, the seventeen-page report of the Committee on Government Operations brought up some very interesting economic points with respect to channelization:

- "Excessive erosion is caused by failure to make proper provisions in the planning of such projects for bank protection and other measures required to stabilize the new channels. The usual reason for omitting these important ancillary measures is to reduce the cost of the channelization project. . . . Had erosion and sedimentation damages been added to the cost of such projects some of them would have failed to meet the test of economic justification."
- "There appears to be a tendency fully to evaluate all benefits that would result from channelization projects, but to underestimate, or even to ignore, some operation and maintenance expenses and damages."
- "Both the Corps and the SCS have often failed to afford adequate opportunity for full public participation in the development of channelization projects."
- "Several large corporations have benefited, or will benefit, from the construction of channelization projects. But these and other beneficiaries often are not identified in the public documents of the Federal agencies doing or financing the channelization work."[34]

How much does channelization cost? Crow Creek on the Tennessee-Alabama line provides an example of cost-benefit ratios. In order to reduce flooding of 125 small farms along twenty-four miles of the creek's floodplain, the Soil Conservation Service straightened, widened, and deepened forty-four miles of Crow Creek in 1971. Parts of the creek had already been widened in the late 1930s. When a team of biologists surveyed the creek a few months after the 1971 work, they pronounced the creek an "ecological disaster"; it had no rooted plants in the streambed or on the eroding banks and no established populations of fish or other animals. Flood protection was estimated in 1972 to enhance the economy of Crow Creek agriculture at $11 per acre per year, but channelization costs were already $1.13 million, which would equal $8.50 per acre per year over the project's estimated fifty-year timespan. Unaccounted for in the cost-benefit analyses were features such as lost habitat, species abundance and diversity, ecosystem functions that promoted clean air and water or soil health, and recreational and esthetic uses of the river corridor.[35]

Published figures on initial construction costs and subsequent maintenance costs, mitigation costs, and indirect losses of natural resources and functions versus flood-control benefits are rare for specific channelization projects. In 1962 it was estimated that damage associated with channel erosion in western Iowa, much of which was initiated by channelization, averaged $719 per square mile. By the end of the twentieth century the region was subjected to $1.1 billion in damage associated with stream erosion. Other examples, including the Demonstration Erosion Control project in northern Mississippi (as discussed later), indicate that the costs of channelization far outweigh the benefits.[36]

Channelization as practiced on a large scale by the federal government declined after the negative attention of the 1970s and the growing realization of the substantial losses to river ecosystems it caused. Channelization did continue, however. A 1993 review of channelization in Illinois noted that nearly a quarter of the state's streams were modified directly by channelization and/or levee construction, generally without effective mitigation. Nationwide, general, and statewide permits continue to be issued for projects that meet specific criteria and guidelines. If the activities involved in the proposed project meet these predetermined requirements, no further evaluation is needed for the work to proceed. As of 2003, the Corps of Engineers was running a nationwide permit system that allowed up to three hundred feet of a small stream to be piped without any site-specific permit. Individual permits are also issued by the corps under Section 404 of the Clean Water Act or the Illinois Department of Transportation. Although permit records may include suggestions for mitigation, permits issued generally do not require even minimal in-stream habitat and bank stabilization efforts. This is in part because numerous projects are permitted after the fact, particularly those on private land. As the 1993 review noted, "institutional constraints, rather than lack of particular understanding about mitigation, provide major barriers to protecting the state's surface water resources." In other words, we as a society know how to protect rivers, but our legal and bureaucratic frameworks often prevent us from using that knowledge.[37]

When environmental mitigation is undertaken during or after channelization, various methods are used. These focus on stabilizing the streambed and banks to prevent erosion and on providing aquatic and riverside habitat. One approach to stabilizing a channel is to control bed erosion and channel downcutting. The streambed can be continuously armored with large boulders. Hard structures can be placed in the channel bed to increase its resistance to erosion. Masonry boulders, concrete, gabions filled with rocks, large wood, and sheet piling are all used to create grade-control structures. These structures are designed to serve as a stable, fixed point along the river's downstream course. The structures send downstream flow over a vertical drop covered in concrete, metal, or stone. A stilling basin in the form of a broad depression in the channel downstream from the drop structure is designed to dissipate the energy released as the flow accelerates over the drop. However, grade-control structures are expensive and prone to failure as scour of the streambed undermines them. They also block fish migration. Beaver are reintroduced to some streams in the hope that they will actively maintain dams that will slow the passage of floods, store sediments, and serve as local grade controls. Such reintroductions succeed where riverside vegetation is sufficient to support the beaver, and where the landowners along the stream channel do not mind occasional overbank flooding associated with beaver ponds.[38]

Ideally, grade-control structures are not placed in a stream without due consideration being given to several factors, including site hydraulics, bank stability, and the potential for overbank flooding. Structures should be placed with regard for environmental impacts, as well as effects on existing structures such as culverts or bridge piers. Likely downstream river response must be considered. The existence of geologic units that may act as grade controls is important when placing structures. Finally, the effects of the structures on tributary streams must be considered. Unfortunately, these factors seldom receive adequate consideration.[39]

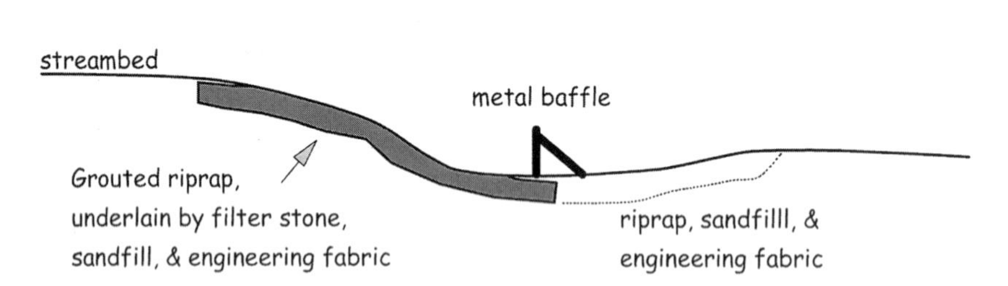

Low-drop grade-control structure with concrete apron downstream, Red Banks Creek, Mississippi, 1995 (top), and schematic side view (bottom). The extensive concrete apron along the streambed and banks below the drop structure is intended to prevent erosion from undermining the drop structure. (Schematic after J. W. Trest, 1997, Design of structures for the Yazoo Basin Demonstration Erosion Control project, in S. S. Y. Wang, E. J. Langendoen, and F. D. Shields, Jr., eds., *Proceedings of the Conference on Management of Landscapes Disturbed by Channel Incision, 1997,* University of Mississippi, Oxford, Figure 2.)

Streambanks are stabilized by covering them with riprap of large boulders, soil cement designed to create a concretelike surface, or gabions or large wood anchored into the banks, or by planting riverside vegetation such as willows. By 1980, channel erosion along Tennessee's Crow Creek created very unstable banks from twelve to twenty feet above the streambed. Stream widths increased from sixty feet to

Rock riprap and scour around bridge piers, Red Banks Creek, Mississippi, 1995. Continuing stream adjustment to channelization has caused severe streambed and bank erosion at this site, threatening the stability of the bridge. The rock riprap is intended to protect the streambank.

more than three hundred feet in some reaches, and sediment deposition downstream was substantial. Between 1987 and 1995, the eroding banks were successfully revegetated with willows. This type of "bioengineering," which is less costly and potentially more environmentally friendly and self-maintaining, increasingly replaces the use of traditional riprap or grade-control structures.[40]

Nonstructural approaches to channel stabilization include increasing the supply of bed load in order to overwhelm the newly energized stream's transport capacity, or excavating a new, lower floodplain so that the stream may adjust within fixed limits.[41]

Mitigation of channelization also focuses on providing habitat without controlling channel erosion. In-channel habitat structures such as rock spur dikes, large wood, or boulder clusters provide stable streambed for invertebrates and localized scour that creates pools for invertebrates and fish. Low dams are used to locally elevate the water surface and reconnect the main channel to abandoned side channels or cutoff meanders, creating backwater habitat.[42]

Longitudinal stone dike with tiebacks, Red Banks Creek, Mississippi, 1995. The dike is intended to reduce bank erosion.

Along the Des Plaines River, mitigation will involve several of these strategies. The stream channel will be regraded and broadened and will be flanked by terraces and experimental wetlands. Stream flow will be pumped to the site perimeter and allowed to gradually return to the river, with sluice gates regulating water passage. Native plant communities will be reintroduced, native fish and wildlife attracted back to the site, and spawning and breeding areas created. Having spent a great deal of time, energy, and money to disrupt a functioning river ecosystem, we now spend even more time, energy, and money trying to re-create what we have destroyed.[43]

Mitigation has been used with some success. One study compared three streams in northwestern Mississippi: downcutting Goodwin Creek, downcutting Bobo Bayou, and stable Toby Tubby Creek. Goodwin Creek was modified to create greater water depth with pools and riffles using small weirs and spur dikes. As a result, mean water depth and pool habitat substantially increased, and mean flow velocity decreased. After rehabilitation, the number of invertebrate species at Goodwin Creek approximated that of the reference stream, Toby Tubby Creek, and exceeded that of downcut and unmitigated Bobo Bayou. Another study focused on five downcut streams to which wood and stone weirs and spur dikes were added. Channel rehabilitation focused on stream reaches downstream of the original channelization that were filling with sediment, with the objective of inducing the formation and maintenance of stable pools. The structures increased pool habitat, fish density, and species richness at some sites but had little effect at others. In some cases, erratic flows were more of a limiting factor for aquatic organisms than restricted habitat.[44]

Any attempt at mitigation of channelization must also consider the stage of channel evolution. Studies in northern Mississippi indicate that eroding streams may benefit from grade-control structures as they approach the stage where they begin to accumulate sediment, whereas channels already filling with sediment do not require such structures.[45]

Numerous studies have been conducted on various aspects of channelization in northern Mississippi, a region that has a long history of river alteration by both local and federal entities.[46]

The DEC Project: Boondoggle or Salvation?

Northern Mississippi has a landscape primed for erosion. For millions of years the Mississippi River and its tributaries have deposited the erosional remnants of two-thirds of the United States across a broad plain here. These weakly cemented sands, silts, and clays are in places overlain by rolling hills formed of loess, the windblown silt deposited along the Mississippi River valley by winds blowing across the vast outwash plains left from the great continental ice sheet of twenty thousand years ago. Onto these highly erodible sediments fall an average of fifty inches of rain a year, most of it during intense thunderstorms. The natural vegetation of forest cover protected the hills and the bottomlands, absorbing the rainfall and releasing it slowly downslope, storing the sediments and the seasonal floodwaters in extensive bottomland hardwood forests and wetlands.[47]

The ecology of the bottomlands centered around seasonal floodings. Each autumn the hardwoods lost their leaves. Bacteria and fungi decomposed the fallen leaves, releasing nutrients that were used by other organisms. Heavy winter and spring rains sent the streams over their banks, and the spreading floodwaters increased the decomposition of the fallen leaves. The dense vegetation of the bottomlands slowed the floodwaters. Suspended sediment settled onto the forest floor, where it combined with the decaying leaves to create rich soil.[48]

Wood ducks and migratory waterfowl such as mallards depended on the bottomland hardwood forests for overwinter habitat. The waterfowl ate acorns, fruits, and berries, as well as the invertebrates found in the shallow waters. These protein sources were essential to reproduction during the subsequent breeding season. Deer, wild turkey, and raccoon also ate heartily of the winter supply of acorns provided by the bottomland hardwood forests. Five acres support a deer on the bottomlands, whereas twenty acres are needed in the uplands.[49]

With the spring warming of the waters covering the floodplain, invertebrates that had overwintered as eggs hatched and began to feed on the bacteria and fungi decomposing the leaves. The waste leaf material expelled by the invertebrates was recolonized by new bacteria and fungi, and a single leaf particle was recycled several times as it provided food for many organisms. Fish also spawned in the overbank areas; bass and sunfish built nests and guarded their young, whereas other species such as shad, buffalo fish, and carp released eggs with an adhesive coating that clung to vegetation. The newly hatched fish then fed on the rapidly increasing numbers of invertebrates in the warm, shallow floodplain waters that provided ideal nursery habitat. As the floodwaters receded back into the river, they carried leaf material that assured a food supply for aquatic organisms in the channel during summer and other periods of low water.[50]

The Native American Chickasaw who lived in the region at the time of European American contact occupied large towns, each of which spread several miles along the riverbanks. The Chickasaw hunted and fished and grew crops centered on maize, beans, and squash. Despite a 1786 treaty with the United States, the Chickasaw were forced to cede portions of their land in 1805, 1816, and 1818 until they were driven into Oklahoma in 1832. European American settlers rushed to occupy the Chickasaw lands in a "speculative mania" that led to widespread cotton farming in the loess hills. By 1860, an observer wrote, "Not only is the soil, and all that could possibly serve as a foundation for the soil, carried away from the hills, but the materials thus removed cover over the fertile branch bottoms, in company with a flood of sand, which renders them useless for all time to come."[51]

The careless use of land by farmers in the southeastern United States was a habit deplored by contemporary observers from other parts of the United States and from Europe. Southern farmers practiced slash-and-burn agriculture, burning and cutting off the native forest cover, planting crops such as cotton for a few years until the topsoil was eroded and the soil fertility gone, and then moving on to another site. The result in an erodible region such as northern Mississippi was the development of a badlands topography with massive erosion from the hills and up to fifteen feet of deposition on the valley-bottom channels and floodplains.[52]

As sediment covered the valley bottoms and choked the stream channels, causing an increase in overbank flooding, local governments organized measures to improve the ability of the streams to convey water downstream. Mississippi enacted its first drainage law in 1886. Two years later the Chiwapa Creek Swamp Land District was created, and this was quickly followed by numerous other districts. These districts were administered by separate boards of commissioners or by the county board of commissioners. As a result of each board's narrow focus, several districts might construct drainage works along a stream without any coordination.[53]

Between 1919 and 1924, cotton prices remained high, and the expanding agricultural economy led to the organization of ninety-five drainage districts incorporating approximately one million acres in Mississippi. Lack of coordination among the districts tended to increase the need for channelization. As an upstream district was channelized, water flowed downstream more rapidly because it was not spreading across floodplains or moving more slowly through secondary stream channels and backwaters. The river segments downstream received higher discharges and then, as the upstream channel downcut and widened, excess sediment. This in turn necessitated channelization of the downstream segments. Included in channelization was channel straightening. Channels with a natural sinuosity of one and one-half to two, measured as the ratio of channel length between two points to straight-line distance between those points, were straightened to a sinuosity of one. This straightening increased the stream's ability to transport sediment by a factor of two to five. Removal of streambank vegetation further increased ability to transport sediment, which rose by fiftyfold relative to natural conditions along some streams. Average annual sediment yield in the region increased from five-hundredths to fifteen-hundredths of a ton per acre per year to fifty tons per acre per year between 1830 and

1920. By the early 1940s, gullies were widespread and upland soil fertility was nearly destroyed.[54]

By the mid-1930s, stream erosion and sedimentation were so problematic that the federal government intervened. The Soil Conservation Service instituted upland soil conservation practices designed to reduce sediment supply. The service also oversaw channel engineering on a watershed basis without restriction by county boundaries. Soil conservation practices included revegetation of critical areas and terracing. Channel engineering included channelization, streambank stabilization, construction of grade-control structures and flood-retention structures in channels, and removal of wood in streams.[55]

The soil conservation practices were largely successful. Soil erosion decreased by an estimated 50 percent between 1937 and 1968. But the channel engineering had mixed results. Channels that during the preceding one hundred years had adjusted to large sediment supplies from the uplands now had a deficit of sediment and began to downcut rapidly. Bank stabilization failed because of channel downcutting that undercut the stabilized banks. Grade-control or drop structures form "fixed points" of resistant material put into the river to stop downcutting. These structures were built in the Mississippi rivers, but they were undermined by aggressive bed erosion. Channel engineering continued to be implemented at a given site without full consideration of upstream or downstream conditions.[56]

By 1985, government scientists working in northern Mississippi recognized that a more integrated approach to channel engineering was needed. The Army Corps of Engineers and the Soil Conservation Service together implemented the Demonstration Erosion Control (DEC) project in the Yazoo River basin of northern Mississippi. This project was intended to develop and demonstrate a watershed approach to mitigating instability associated with erosion, sedimentation, flooding, and environmental degradation. DEC activities focus on fifteen watersheds comprising twenty-three hundred square miles in the Yazoo River basin. Here, the suspended sediment yields run about ten times the national average for similar-sized watersheds, despite declining to five to ten tons per acre per year by 1980. To address these channel and landscape instabilities, DEC watersheds were modified in several ways.[57]

The DEC project is worthy of federal-level intervention in its intensity. The 2,300 square miles of the Yazoo basin include more than

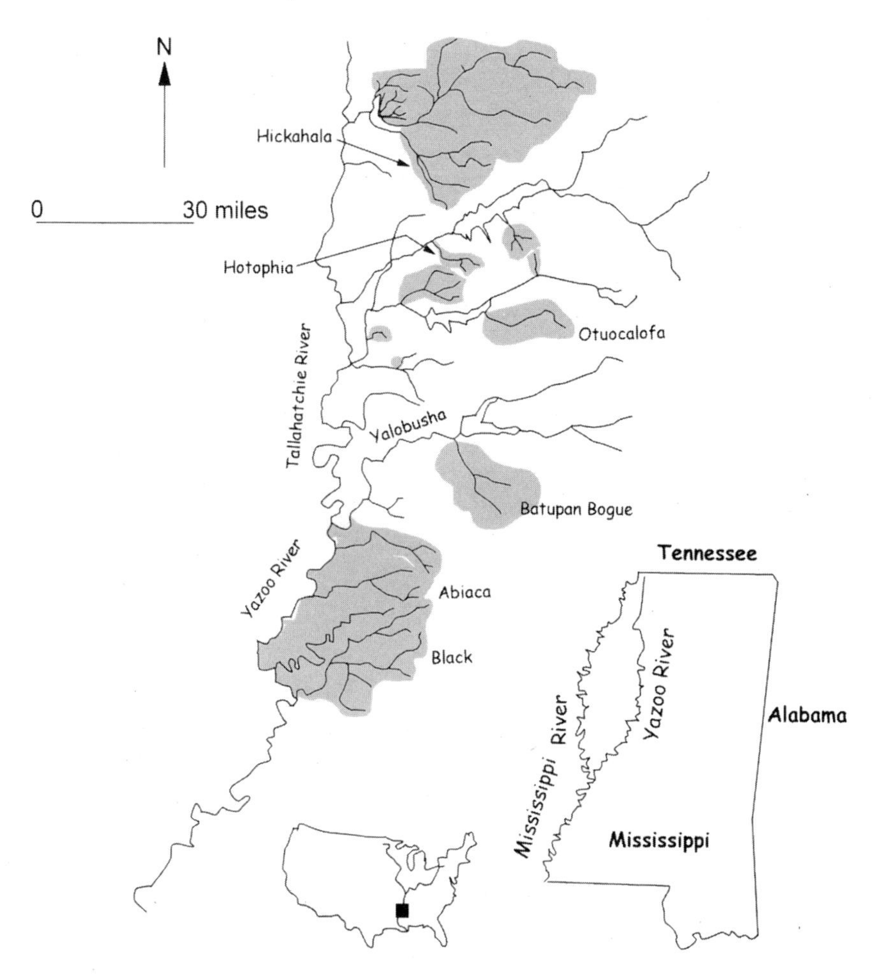

Location of the Demonstration Erosion Control project area. (After F. D. Shields, Jr., S. S. Knight, and C. M. Cooper, 1995, Rehabilitation of watersheds with incising channels, *Water Resources Bulletin*, 31, 971–82, Figure 1.)

twenty-three hundred grade-control structures, seventy-two floodwater-retarding structures, more than two hundred debris basins, many miles of levees, and about 310 miles of bank stabilization. Perhaps most importantly, the DEC project includes long-term monitoring of structural performance, habitat enhancement, and overall system response to channel engineering as measured in sediment yield. On this last criterion, the project appears to be succeeding. Sediment yield in the monitoring reaches decreased by 22 percent from 1992 to 1995.[58]

Reaching this point of relative success has been problematic. True to

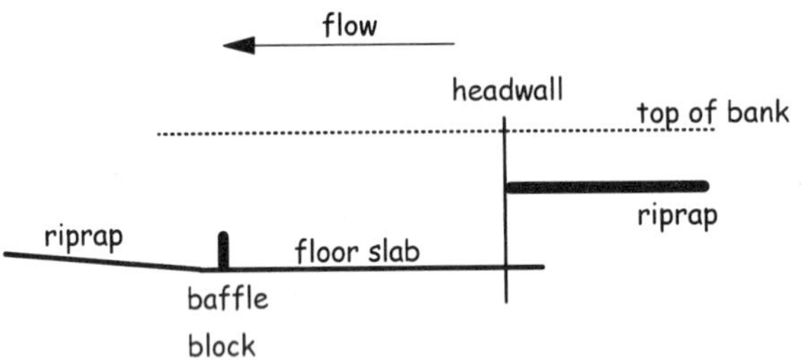

Top: View upstream to a high-drop grade-control structure built by the Soil Conservation Service along Burney Creek, Mississippi, 1995. Bottom: Schematic side view of a high-drop grade-control structure. The channel banks above and below the headwall are also protected with riprap. (Schematic after J. W. Trest, 1997, Design of structures for the Yazoo Basin Demonstration Erosion Control project, in S. S. Y. Wang, E. J. Langendoen, and F. D. Shields, Jr., eds., *Proceedings of the Conference on Management of Landscapes Disturbed by Channel Incision, 1997,* University of Mississippi, Oxford, Figure 3.)

the tradition of channel engineering, initial structures were placed without regard for upstream and downstream river conditions and without attention to long-term monitoring. As a result, many of these structures failed. However, lessons learned from these failures were incorporated into subsequent structural designs. Of fifty-five low-drop structures averaging 2.5 years of age that were inspected in 1993, 9 percent were rated to be near failure. Of seventy-six structures averaging 3.8 years of age inspected in 1996, all were considered functional. This in-

Bank-protection structures along Burney Creek, Mississippi, 1995.

creased success represents a process of fine-tuning the design of the structures by changing their dimensions and the size of the riprap used to construct them.[59]

Structures sometimes also exacerbated erosion elsewhere along the channel. The lower eleven miles of Hotophia Creek were channelized in the early 1960s to reduce the frequency of floodplain flooding. The channel then began to erode its bed and banks. Fourteen grade-control structures were built along the channel between 1980 and 1996 in order to aid channel recovery and reduce erosion rates. These structures served their purpose by inducing sediment deposition and reducing the migration of headcuts upstream of the structures. By reducing the downstream sediment supply, however, the structures also enhanced streambed and bank erosion downstream.[60]

Although channel evolution models such as the six-stage model dis-

Burney Creek, Mississippi, 1995, completely contained in a concrete canal where it passes through a suburban area.

cussed earlier are used to evaluate whether grade-control structures or other stabilization measures are appropriate, these models provide approximations of channel response through time, rather than precise predictions. Equations relate factors such as drainage area to the width of filling or downcutting channels, but variability among individual sites limits the precision of these equations. Extensive mapping of regional geology and soils in relation to bank stability indicates that river form is partly controlled by the nature and distribution of specific sediment units, which range from consolidated sands and gravels through cohesive silt and clay to weakly consolidated sands.[61]

Similar to the learning process associated with the design of drop structures, bank-protection measures were installed based on professional judgment, rather than on specific guidelines for location and

View of older, low-drop structure with baffle plate, Hotophia Creek, Mississippi, 1995.

angle, because such guidelines did not exist. An analysis of a bank-stabilization project along Harland Creek, for example, indicated that thirty-five of the fifty-four bendway weirs emplaced along 2.2 miles of the creek in 1993 were initially located and angled incorrectly. Bendway weirs are structures built perpendicular to the streambank in order to direct erosive flow away from the bank. Postconstruction monitoring of the $323,000 project resulted in changing the location and angle of some of the weirs, and in improved guidelines for future weir installation.[62]

The DEC project has not been inexpensive. Low-drop structures typically cost $200,000 to $400,000, depending on the size of the stream. As of 1995, estimated total costs for the DEC project were $526 million, an average of $246,900 per square mile.[63]

In addition to the obvious costs of construction and monitoring, there are unmeasured costs of environmental degradation. In general, stream fish assemblages are limited by lack of suitable habitats in the DEC region. Estimates are that federal flood-control work dating from 1939 onward reduced stream reaches capable of supporting any kind of fish by 80 percent. Pool habitat, wood, and riverside vegetation are scarce. Pools are particularly transitory, eroding or filling repeatedly as flow changes. Low-flow depths are shallow and floods are brief. Fewer

Aerial view of grade-control structure along Hotophia Creek, Mississippi.

than 10 percent of the channelized streams sampled during low flow occupied more than 5 percent of their channel cross sections. For comparison, nonincised streams occupy an average of 20 percent of the channel cross section during low flow.[64]

Estimates are that before federal channelization works, the Yazoo basin provided more than five million pounds of fish per year. When channelization reduced overbank flooding, fish that spawned in the floodplain forests tried to reproduce in the channelized streams. The loss of in-channel vegetation and wood meant that the fish eggs settled to the streambed, where they were covered by sediment and smothered. The newly hatched juvenile fish were susceptible to death through collision with sediment particles no more than an inch in size. Where bottomlands occasionally flooded, fish attempting to spawn found agricultural lands barren of vegetation and food but permeated with chemicals. Analyses of fish collected from the Yazoo River since 1970 indicate extremely high concentrations of more than twenty pesticides and pesticide derivatives in the fish tissue. In unchannelized stream segments with intact bottomland forests, estimates of commercial fishery standing stock range from 150 to 300 pounds of fish per acre. By 1960, standing stock in the Yazoo basin declined to 24.5 pounds per acre, and by

1970 to 11.3 pounds per acre, with projected continuing declines of 10 percent per year.[65]

Streambeds in channelized systems are dominated by mobile sands. Because the species diversity of bottom-dwelling insects declines as streambeds become more mobile, the sand-bed channels have invertebrate communities characterized by few species. Channelized streams convey higher levels of nutrients and suspended sediments than non-incised channels with forested wetland floodplains, particularly during and immediately after rainstorms. The loss of the riverside buffer zone also results in higher levels of streambed contamination with arsenic, mercury, and residues from organochlorine pesticides.[66]

More than 75 percent of the original acreage of bottomland hardwoods in the Yazoo basin was lost to erosion or agriculture. This loss of overwinter feeding grounds affected mallard and black ducks, ringnecks, green-winged teal, gadwalls, blue-winged teal, and hooded mergansers. The wood ducks who feed and nest in the bottomland forests were also strongly affected. The deer lost the carbohydrate-rich acorns that built the layer of fat they needed to see them through the winter. Forest birds, squirrels, mink, otter, beaver, bald eagle, black bear, and alligator lost vital floodplain habitat.[67]

In some cases, structures designed primarily to stabilize channels also improve aquatic habitat. Channels with drop structures and associated stilling-basin pools support a distinct and more diverse aquatic community than downcut channels without drop structures. The drop structures can be modified to create more aquatic habitat by increasing the overall hydraulic complexity of physical habitats. Stony riprap associated with drop structures and bank stabilization provides stable surfaces for colonization by macroinvertebrates, replacing some of the habitat provided by wood in non-downcut channels. The riprap habitat is not as good as that provided by wood, however, and this is another reason that planting of sapling trees is increasingly used to stabilize streambanks. Where levees or dams have ponded water away from the river channel, a drop pipe may be installed to siphon water from the pond to the river and maintain water level in the pond. Drop-pipe habitats are used by amphibians, and the wetlands created upslope from drop pipes are heavily used by fish, amphibians, reptiles, birds, and mammals. Although fish populations in some mitigated streams partially recover in terms of density, the species composition of the community

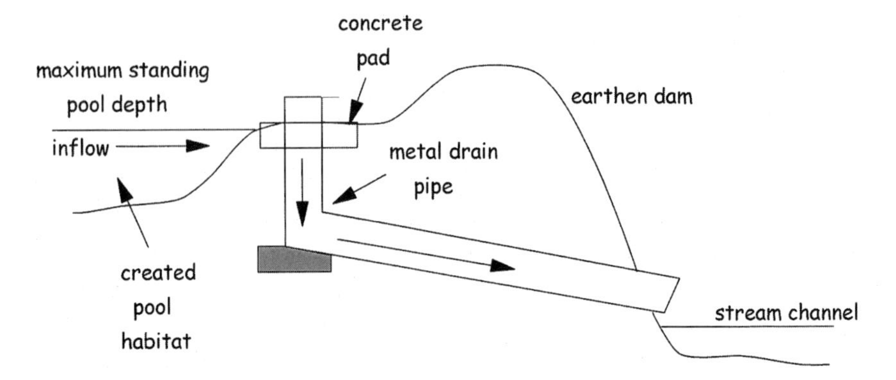

Schematic cross section through a typical drop-pipe structure and created habitat. (After P. C. Smiley, Jr., C. M. Cooper, K. W. Kallies, and S. S. Knight, 1997, Assessing habitats created by installation of drop pipes, in S. S. Y. Wang, E. J. Langendoen, and F. D. Shields, Jr., eds., *Proceedings of the Conference on Management of Landscapes Disturbed by Channel Incision, 1997,* University of Mississippi, Oxford, Figure 1.)

shifts from species such as shiner, madtom catfish, and darter to more tolerant species such as shad, bluegill, mosquito fish, and bluntnose minnow.[68]

The "replacement" of one fish species by another is worth considering closely. Not only are the replaced species akin to a miner's canary signaling a dangerous change in the ecosystem, but these species also represent a tragic loss in their own right. The southeastern United States has the highest diversity of fish species in the country. After hundreds of millions of years of evolutionary adaptation to the rivers of this region, these fish are vanishing in a few decades of human manipulation and exploitation. The paddlefish is one of the saddest losses.[69]

A Beaver-Tailed Fish

The Mississippi is a giant among rivers. The flow of the river ranks as the world's third largest, behind only those of the Amazon and the Congo. The Mississippi's drainage basin covers 41 percent of the contiguous United States, a huge area of 1,152,000 square miles. From this area the river annually transports an average of 750 million tons of sand, silt, and clay. During the flood year of 1993 the river carried 2.1 billion tons of sediment into the Gulf of Mexico. These sediment loads keep the river

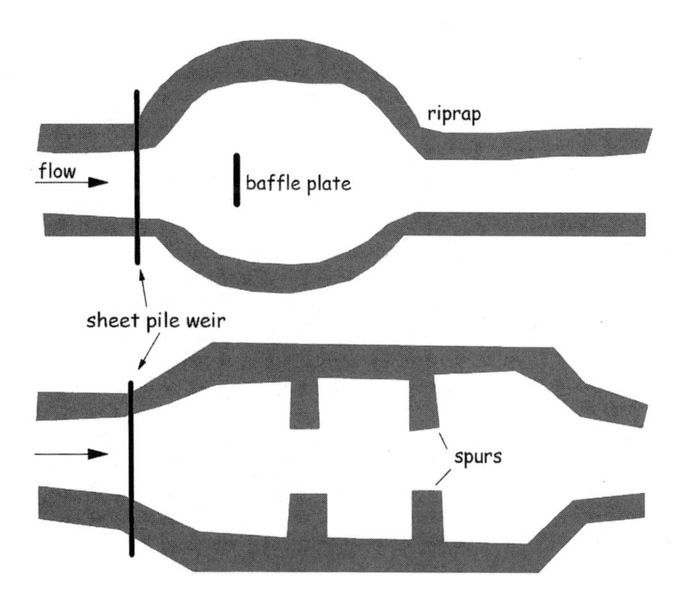

Plan view of a standard low-drop grade-control structure (upper) and structure modified to improve aquatic habitat by removing the baffle plate, lengthening the basin, and adding spurs (lower). (After F. D. Shields, Jr., S. S. Knight, and C. M. Cooper, 1995, Rehabilitation of watersheds with incising channels, *Water Resources Bulletin,* 31, 971–82, Figure 4.)

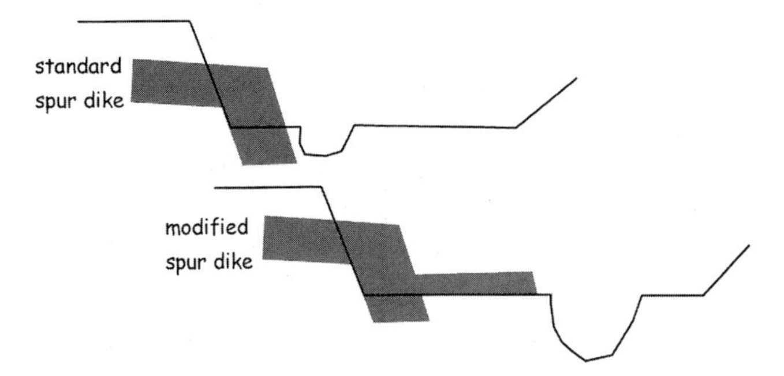

Schematic cross-sectional view of a standard spur dike design used for bank stabilization and resultant scour pattern (upper) and a modified design with an enlarged scour hole and pool habitat (lower). (After F. D. Shields, Jr., S. S. Knight, and C. M. Cooper, 1995, Rehabilitation of watersheds with incising channels, *Water Resources Bulletin,* 31, 971–82, Figure 6.)

Drawing of a paddlefish. (Courtesy of Joe Tomellerie.)

muddy; "Big Muddy" it has been called, as well as "Father of the Waters." A fish needs special skills to find its way about in these muddy waters, and the paddlefish possesses these skills.[70]

The paddlefish is a giant among river fish. Historically, these fish grew to at least six feet in length and two hundred pounds in weight. They can live at least thirty years, feeding almost constantly and continuing to grow as long as they live. And they are prodigious swimmers. Paddlefish travel five hundred miles to spawn, and they swim along at three miles an hour against a current running at the same rate, with occasional bursts of speed up to ten miles an hour. One fish moved nearly twelve hundred miles within eight months of being tagged.[71]

Paddlefish (*Polyodon spathula*) are often referred to as primitive fish because they belong to one of the two families of fish that have survived largely unchanged in North America since the Cretaceous Period 140 million years ago. During this long span of time, the paddlefish spread throughout the Mississippi River and its major tributaries, from Montana to West Virginia and from the Great Lakes down to Louisiana and into smaller rivers draining to the Gulf of Mexico.[72]

These diverse rivers provided good habitat for the paddlefish. The big fish evolved a distinctive means of gleaning food from the surrounding murky waters. Up to a third of the fish's body length is a rigid, broad, flattened snout slightly reminiscent of a beaver's tail. This snout contains tens of thousands of ampullae of Lorenzini, tiny electroreceptors that detect the very small electrical signals generated by zooplankton such as *Daphnia*, water fleas only a fraction of an inch in size. These tiny fleas and other minute water creatures are the prey of the big fish, which cruise along with their mouths open. When open, an adult paddlefish's mouth encompasses an area about the size of a gallon bucket. Sievelike gill rakers within the mouth filter the small food particles from the water

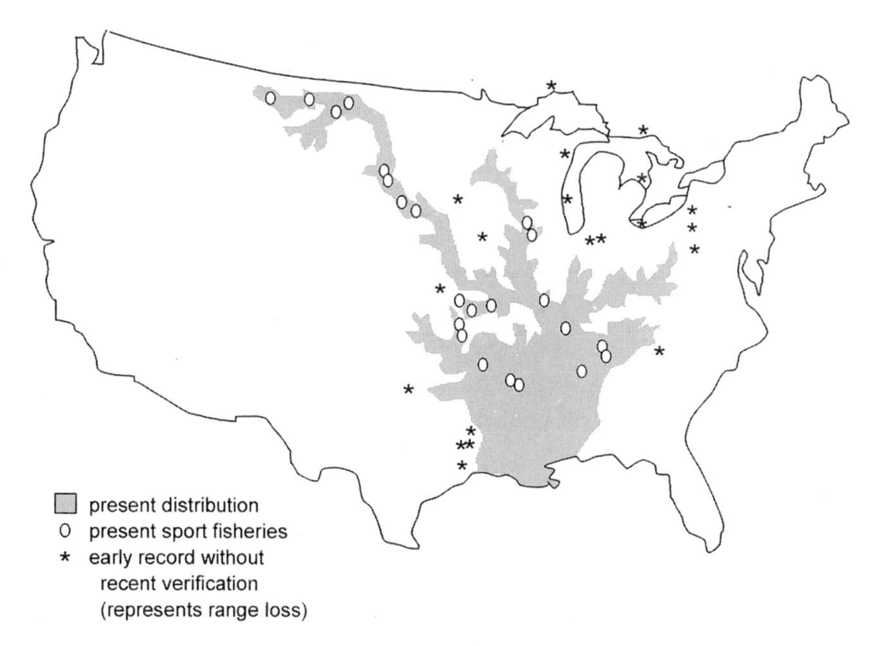

Past and present distribution of the paddlefish and locations of sport fisheries. (After J. G. Dillard, L. K. Graham, and T. R. Russell, eds., 1986, *The paddlefish: Status, management, and propagation,* Am. Fisheries Soc. Special Publ. No. 7.)

and suspended sediment. The tiny eyes of the paddlefish have little to do with finding food or avoiding obstacles.[73]

Adult paddlefish move upstream to spawning areas when the river clues them in that spring has arrived. Once the river flow rises several feet, and the water warms to 50 degrees Fahrenheit or more, the fish begin to move. The adults continue upstream until they find cobble and gravel bars clean of silt and clay. Here, the females release their eggs in a series of groups over twenty-four hours or longer. If conditions are not right, the females resorb the eggs and retreat downstream without spawning. The females spawn at intervals of two years or more and do not reach sexual maturity for six to twelve years. Thus, a few missed spawning seasons seriously affect a population.[74]

The eggs are adhesive and become firmly attached to the streambed gravel. With water temperatures around 57 degrees Fahrenheit, the eggs hatch in ten to eleven days. Within a month of hatching, the larvae drift or swim into areas of low flow velocity with abundant zooplankton and begin to feed. The young paddlefish grow rapidly—from less than an inch in April to twenty-eight inches long by October.[75]

When they are not spawning, paddlefish inhabit relatively deep, slow- or still-water areas with flow velocities of less than a foot per second. Backwaters, or main channel borders downstream from backwaters or obstacles such as bars, provide reduced flow velocity and a drift of zooplankton on which the paddlefish can feed.[76]

Not much feeds on a big paddlefish. And the paddlefish's prey and feeding behavior minimize competition with other species. But the giant fish, once one of the most abundant species in the Mississippi River drainage, is today listed as a species of concern in much of its range and is largely extinct in the eastern half of its former range. During the late 1800s, fishers discovered that a big female paddlefish holds up to twenty-five pounds of eggs in the spring. These eggs became a domestic source of caviar, and by the 1890s two million pounds of paddlefish were taken annually in the Mississippi, Ohio, and Missouri Rivers. The catch was unregulated and had declined drastically by the time of the first formal study of paddlefish in 1907. At that time, paddlefish were already extirpated in Maryland, New York, North Carolina, Pennsylvania, and Canada. There have not been enough large paddlefish to support a commercial fishery since 1907 except locally in reservoirs, and these fisheries were also overexploited by 1980. Fisheries biologists estimated that by the 1980s the Mississippi River drainage system supported only 10 to 20 percent of the paddlefish numbers once present.[77]

Overfishing severely stressed paddlefish populations, but habitat loss is commonly identified as the most important factor in the species' decline. Spawning areas were destroyed by siltation under the ponded water of reservoirs, or by erosion during higher-than-normal peak flows and subsequent exposure during very low flows along channelized rivers. Dams reduced upstream and downstream movement to spawning areas that still exist. The finely tuned electroreceptors that detect swarms of tiny *Daphnia* are more than adequate to detect the large metallic structures of locks and dams. The paddlefish responds to this detection with avoidance, making it difficult for the fish to swim upstream or downstream through these structures. Channelization largely eliminated the backwater areas so important to both young and adult paddlefish. It also eliminated habitat diversity created by bars or islands. All but 18 of the 161 major islands in the Missouri River were removed, and with them went the zones of low velocity where paddlefish feed on zooplankton drifting downstream. The dewatering of channelized

Top: Aerial view of a channelized reach with grade-control structure, Indian Creek, Mississippi. The original, meandering channel is the wooded belt to the right. Left: An unchannelized stream, Coldwater Creek, Mississippi. The difference between the form of this stream and that of the channelized stream illustrates the type of habitat complexity that is critical to paddlefish survival but that is lost as a result of channelization.

streams associated with shorter duration, larger flows desiccates paddlefish eggs. The passage of a large commercial vessel creates a two- to three-minute-long recession in water level along a river shoreline, and this can strand young paddlefish. Turbulence from the vessel's passage may also kill paddlefish larvae.[78]

Paddlefish adapt to these changes as they can. They use tailwater habitats in pools associated with locks and dams. They congregate near structures such as wing dikes, which create eddies and reduced flow velocities. But there is clearly a lack of suitable habitat. A 1992 study of paddlefish movement in Pool 13 of the upper Mississippi River found that three-quarters of all contacts with paddlefish occurred in about 5 percent of the available seven types of habitat at the site.[79]

Pollution has further stressed the paddlefish. PCBs are lipophilic; they concentrate in tissue with a high lipid, or fat, content. In the paddlefish, this tissue is skin, red muscle, reproductive tissues, and eggs. The paddlefish take up PCBs directly from the water through the gills and other body surfaces, or from food consumed, or from the fine sediments the fish ingests along with the zooplankton. One study found PCB concentrations of sixteen to forty-three parts per million in paddlefish eggs from the Ohio River near Louisville, Kentucky. These may seem vanishingly small amounts, but PCBs are so toxic that these concentrations signal health hazards for the young paddlefish. Beneath its tough, scaleless hide, the paddlefish has a thick layer of red flesh laced with blood vessels. This layer helps the fish to absorb oxygen in deep, slow-moving water, and it helps the fish to be a stalwart swimmer. But this layer of red muscle also concentrates PCBs at significantly higher levels than white muscle tissue.[80]

What may help to save the paddlefish, by bringing human attention to its plight, is our continuing fondness for paddlefish as food and sport. Fishers discovered in the mid-1930s that snagging paddlefish by dragging big triple hooks through the channels where the fish congregated made good sport. This, combined with the taste of paddlefish white meat and roe, has kept fisheries managers focused on the fish. Spawning habitat has been protected in the Upper Missouri and Yellowstone Rivers, and hatchery programs have met with some success. Yet as of 1986, most harvest regulations were guesswork. Fourteen of twenty-two states regulated sport or commercial harvest of paddlefish, yet only four states reported that the effects of their regulations were known. And the rise in caviar prices during the 1980s and 1990s was not good news for the sturgeon and paddlefish, the sources of U.S. domestic caviar.[81]

The paddlefish is something of an anomaly among river fish. It has no scales and no bone; its big body is supported solely by cartilage. Its long, flat snout results in the nickname of spoonbill catfish. Its tiny eyes are of little use, but the numerous electroreceptors growing in its flesh can detect minuscule water organisms. It can synthesize ascorbic acid, an ability many higher vertebrates such as humans have lost; we get scurvy without an external source of the acid. At the start of the twenty-first century, the paddlefish is a living fossil, an evolutionary curiosity. Yet it is also a species superbly adapted to exploit the riches of North America's largest river system, as well as a representative of the stresses imposed on thousands of other species by our heedless attempts to make rivers conform to the most restricted and short-sighted vision of human needs. As we strive to undo some of the damage we have caused, and to rehabilitate and restore river ecosystems, our criterion for success is the ability to mimic a functioning river that can support the paddlefish and all the wondrous diversity of riverine organisms for generations to come.[82]

Chapter 6

Trying to Do the Right Thing

Rehabilitation Impacts

More than six billion humans inhabited the Earth at the start of the twenty-first century. This single fact is the best indicator of our short-term technological successes because it implies a tremendous ability to alter the planet in order to facilitate immediate human needs. In the past, various human societies were overwhelmed by environmental crises that they did not have the technology to overcome. The Salado culture of central Arizona abandoned agricultural fields rendered infertile by salinization associated with irrigation during the fourteenth century. The city-state of Mohenjo-daro was abandoned around four thousand years ago, apparently in response to a flood or series of floods on the Indus River that destroyed the city and its surrounding fields. The Norse settlers in Greenland slowly lost their vegetables and hayfields, and their ability to live in the region, as climate became progressively colder during the twelfth and thirteenth centuries. Each of these cultures was destroyed or altered dramatically by an environmental change, but each had the potential safety valve of a new physical frontier to colonize. As humans continue to develop new technologies, we increasingly insulate ourselves against localized natural disasters such as floods or droughts, as well as against human-induced disasters such as soil salinization. Yet we do not fully insulate ourselves, and we have no remaining frontiers. This is a crucial difference between historical and contemporary examples of environmental crises. A large measure of the limited insulation that we do have derives from landscape engineering.[1]

Landscape engineering—the alteration of the surrounding environment to meet human needs—has a long history. As early as 2800 B.C., Egyptians dammed their rivers to reduce floods and store water for human use. The early Japanese began reconfiguring the mountainous landscapes of their islands as early as ten thousand years ago. They filled and terraced coastal wetlands to create farm fields, and terraced the mountainous slopes to reduce debris flows. Li Ping initiated an extensive system of irrigation canals and flood-control structures more than twenty-one hundred years ago in the Szechwan region of China. In A.D. 1750 the Rio Guadalquivir in Spain was channelized, reducing its length by 40 percent. By that time, levees were already built along the lower Mississippi River to reduce overbank flooding.[2]

As human population density and technical ability increased, the extent and intensity of landscape engineering increased proportionally. In the former Soviet Union, the waters of two of Central Asia's major rivers were diverted from the Aral Sea to nearly twenty million acres of land devoted to irrigated agriculture. Between 1960 and 1995, the Aral Sea lost two-thirds of its volume, and salinization began to affect farm fields and groundwater supplies. During 1990 alone, approximately forty-two hundred companies mined nearly one billion tons of sand and gravel from fifty-seven hundred operations along rivers and floodplains in the United States. This mining of construction aggregate causes channel erosion, downstream siltation and reduced water quality, lowered water tables, and destruction of aquatic and riparian habitat. In the Appalachians, mountaintop removal of coal continued apace as of 2004. This mining involved removing up to six or seven hundred feet of material above a coal unit, and simply dumping that material into the nearest valley. More than 900 miles of streams throughout the eastern coal states were buried in this manner. West Virginia alone has 470 miles of obliterated streams in five of the thirteen coal counties that were checked by the Fish and Wildlife Service. This activity is better termed landscape obliteration, not landscape engineering. The narrowly averted climax of this arrogant attitude toward landscape modification was Edward Teller's proposal in the 1950s for Project Chariot, in which six thermonuclear bombs would be detonated to create a harbor above the Arctic Circle in Alaska. Only courageous work by biologists and Native Americans and their allies prevented a project that would have contaminated the Arctic food web with radioactive fallout.[3]

Whether accidentally or deliberately, every landscape on Earth is now altered by humans to some extent. Considering only the rate at which humans move earth materials, our role in shaping the landscape is greater than that of any geologic agent. One geologist has estimated that the total earth moved by humans during the past five thousand years is sufficient to construct a mountain range thirteen thousand feet high, twenty miles wide, and sixty miles long. Human activities annually displace approximately thirty-five billion tons of earth, whereas rivers transport approximately twenty-four billion tons a year, of which ten billion tons are directly attributable to agricultural activities. Humans alter topography, hydrology, biology, biochemistry, and even climate. Our activities are usually not so much engineering—which implies a rational approach with carefully designed controls to eliminate side effects or unforeseen consequences—as blind tinkering. The often horrifying unforeseen consequences of our actions do not stop us from tinkering, but in some cases that tinkering takes the form of trying to restore or rehabilitate some of the ecosystem functions that we have destroyed.[4]

When applied to rivers, restoration is strictly defined as a return to a close approximation of the river condition before disturbance. Rehabilitation refers to improvements in condition that do not attempt any return to predisturbance conditions. Rehabilitation is sometimes described as putting the channel back into good condition. Restoration includes both river form and function, whereas rehabilitation may focus on river form. However, both terms—restoration and rehabilitation—are loosely applied to a wide range of activities.[5]

River rehabilitation is undertaken for many reasons. Sometimes the intent of rehabilitation is to reduce channel erosion in order to improve water quality or control eroding banks threatening structures or causing property loss. Sometimes the intent is primarily cosmetic. If it is appropriate to speak of river fashions, the multiple, rapidly shifting channels of braided rivers are unfashionable, and meandering rivers with pools and riffles are fashionable. Too often, we try to force naturally braided rivers into a single meandering channel deemed more suitable for a trout fishery.

Often the intent of a river rehabilitation project is to improve aquatic or riverside habitat for a particular species. This may present special challenges in that it is difficult to quantify the habitat needs of a species. Many species of fish, for example, need pools for overwinter and resting habitat, riffle gravels for spawning, and shallow, low-velocity areas such as floodplains or backwaters for nursery habitat. But just how deep must the pools be, and for how long do the fish need flooded backwaters? Biologists address these questions using habitat suitability curves that assign numerical values to such habitat characteristics as grain size on the streambed or flow depth. Development of the habitat suitability curves requires prolonged, careful study of the life history and habitat use of each species. In addition, fish rely on a food web of organisms, each of which has its own habitat needs.[6]

Rivers are restored or rehabilitated using various technologies, although some of these do not meet the strict definition of restoration adopted by the National Research Council's Committee on the Restoration of Aquatic Ecosystems. Constructed wetlands or a buffer strip of vegetation planted along the river corridor reduce the levels of sediment and other contaminants entering the river, stabilize the stream banks, redirect the current to promote meandering, and provide shade and organic matter input for aquatic organisms. Traditional engineering materials such as rock riprap or soil cement, or bioengineering methods such as planting willow posts or anchoring logs to the banks, directly stabilize streambanks. The river form is reengineered, with earthmoving equipment used to contour meanders, pools, and riffles. Gravel and cobbles are emplaced on the streambed, or isolated logs or boulders are used to cause localized scour of the streambed and the formation of pools. Weirs are installed to pool water upstream. Beaver are reintroduced to the river. Levees are intentionally breached to restore connections between the river and the floodplain.[7]

River function may be assisted by preserving in-stream or channel-maintenance flows. In-stream flow regulations generally mandate some minimum discharge within the stream channel to preserve such qualities as fish habitat or recreational uses. Channel-maintenance flow regulations focus on the volume of flow necessary to flush sediment through the channel and maintain the channel's ability to convey water downstream without an increase in overbank flooding. In the absence of these regulations, many rivers in the arid and semiarid western United States are literally drained dry for off-stream agricultural and municipal water uses.[8]

Dams constitute one of the more severe and widespread human impacts on U.S. rivers, and one of the more challenging scenarios under which river rehabilitation may be undertaken. The physical and biological disruption of rivers dammed from the 1940s through the 1970s was apparent by the 1980s. National social values had shifted toward a greater emphasis on environmental quality, and federal legislation protecting endangered species had created a legal framework for habitat protection and rehabilitation. In this setting, people began to question whether some of the nation's more than seventy-five thousand dams should be removed, or at least operated so as to mimic the timing and changes in magnitude of natural flows. As of 2003, experiments in mimicking natural annual flow patterns were underway in the Mississippi and Missouri River systems and on the Rio Grande, Colorado, Kissimmee, and Trinity Rivers.[9]

These experiments, or proposals to conduct them, generate stiff opposition from dam operators reluctant to lose water stored for dry periods or revenue from hydropower generation. Utilities using hydropower, farmers relying on water supplies, or businesses relying on dam-regulated flows for transport also object. In early 2002 the National Academy of Sciences issued a report calling for more flow releases in spring and fewer in summer along the Missouri River, as well as reactivation of some of the river's natural meanders. Midwestern farmers and barge operators immediately protested, and their congressional representatives echoed the protests. The environmental group American Rivers named the Missouri America's most endangered river in 2002, largely because of environmental losses associated with dams.[10]

Dam removal is equally difficult to implement. The majority of the dams removed in the United States as of 2002 were fewer than thirty feet tall. Even for these relatively small dams, removal is not a simple physical process of breaching concrete or stone. Residents near the dam site may oppose dam removal for economic, esthetic, or safety reasons if they use the reservoir for recreation or fear increased flood hazards after the dam is removed. Removal of a dam results in greatly increased sediment transport downstream as reservoir sediments are mobilized. These newly exposed sediments can serve as germination sites for weeds or exotic riverside vegetation. Sediment mobilization can also disperse contaminants such as organochlorine compounds or metals that are associated with the sediments. And large pulses of sediment can alter downstream habitat by covering spawning gravels, filling pools, changing water depth and temperature, reducing exchange between the stream channel and the underlying hyporheic zone, and loading the stream with nutrients. Successful removal of any dam requires careful planning and control of the rate and manner of channel adjustment to the new discharges of water and sediment.[11]

New attempts to restore or rehabilitate rivers have become increasingly widespread in Europe and the United States since the 1970s, but we have much damage to undo. In the United States, we have about 3.2 million miles of rivers. Seventy-nine percent of these river miles are somehow affected by human activities, and another 19 percent are drowned by reservoirs, leaving only 2 percent of relatively unimpacted rivers. Most people living in the United States have probably never seen a river unimpacted by humans. However, as we attempt to "fix" our rivers, we are learning that we have much yet to learn.[12]

Nothing New Under the Sun

Despite the recent impetus for river rehabilitation, the concept of "improving" rivers has a long history in the United States, as traced by Douglas Thompson in a series of papers. Increased angling and decline in fish numbers led to the privatization of stream reaches in the Catskills region of New York shortly after the Civil War. Privatization was augmented by fish stocking and by attempts at enhancing fish habitat. These enhancements included low dams built of boulders, logs, or concrete. The dams were designed to create pools by ponding water upstream and increasing streambed scour where flow plunged over the dam. Other enhancements took the form of devices to deflect stream flow and create streambed scour, or structures to create cover and shelter for fish. Emplacement of these various structures by wealthy sport fishers grew increasingly widespread between the 1890s and 1930, particularly in the northeastern United States.[13]

Government and academic research on the use of structures to improve fish habitat began in Michigan during the early 1930s. The motivation was the desire to shorten the time between bites for anglers, and to reduce the long walks between successive pools so that "fishing would bear less resemblance to golf." Fish scientists such as Carl Hubbs believed that streams could be modified to almost any degree desired. This philosophy was adapted by the Civilian Conservation Corps (CCC) when it began stream improvement in 1933. Within a year, the CCC initiated projects on public lands from Arizona to New York and

Michigan to West Virginia. Between 1933 and 1935, the CCC constructed 31,084 structures on 406 mountain rivers. By 1936 the busy workers "improved" 4,770 miles of streams. These improvements, which were considered experimental, were sometimes placed without a preliminary survey of the channel by workers who came from urban areas and had never before seen a trout stream.[14]

The Forest Service produced its first handbook on the use of habitat structures in 1936. The designs in this manual changed remarkably little in succeeding decades, and users conducted few comprehensive evaluations of the effectiveness of the structures until the 1990s. These recent evaluations indicate that the majority of structures are ineffective or even detrimental because they continue to represent static designs that attempt to impose an effect on a dynamic river system, or because they are based on inappropriate assumptions regarding river behavior.[15]

Several examples illustrate these shortcomings. Modern design standards recommend the use of low-profile flow deflectors. These are intended to enhance pool formation, primarily during lower flows. Low profiles are preferred to minimize protrusions on which drifting debris can accumulate and to guide stream flow rather than dam it. However, pools are typically formed during higher flows, which tend to drown out low-profile deflectors and render them inactive. A 1990s study of low deflectors installed along Connecticut's Blackledge River during the 1930s indicated that these deflectors create only minimal pool habitat. Cover structures installed along the same river retard the growth of streamside vegetation by a third relative to unaltered reaches of the stream, and this in turn causes 75 percent less overhead cover on "improved" stream reaches than on unaltered reaches. Stream relocation during "improvement" of another portion of the Blackledge River during the 1950s resulted in overbank erosion and formation of a cutoff channel because the processes controlling sediment deposition along the river were ignored. Similarly, a newly constructed meandering portion of Uvas Creek in California was destroyed during a 1996 flood, only a few months after channel reconstruction was completed. The reconstructed channel was designed on the basis of assumptions that a meandering gravel-bed stream would be stable at the site and that channel form was determined by flows with a return period of about one and one-half years. Both of these assumptions proved incorrect. The recon-structured channel ultimately failed because a new channel form was imposed without addressing the processes that determine channel form.[16]

Reconsidering Earlier Assumptions

One of the most difficult concepts to apply in river restoration or rehabilitation is the idea that each river is a uniquely functioning ecosystem that is dynamic in space and time. Many of those attempting river restoration today rely too heavily on computer models, without field measurements or observations. As geomorphologist Luna Leopold wrote: "In river work, computer modeling is an insidious procedure in which an air of surety hides questionable assumptions." Failure to realize that each river has unique aspects results in projects focused on a single river reach, without sufficient consideration of upstream-downstream or river-hillslope interactions. Rehabilitation that creates new pools downstream from a rapidly downcutting stream segment, for example, is likely to result in infilling of the new pools unless the excess sediment being carried downstream from the zone of downcutting is somehow trapped above, or routed through, the rehabilitated reach.[17]

Rehabilitation in isolation can also result in direct damage to other sites. Hillslope or streambank stabilization sometimes utilizes topsoil strip-mined from somewhere else. Instead, the soil existing at the site can be amended with nutrients, composted yard waste, or soil fungi, as appropriate. Vegetation native to the site, or tolerant of soils with mineral imbalances, can be planted to further stabilization. Localized streambank stabilization can also enhance downstream bank erosion when energetic floodwaters that were deprived of sediment while flowing through stabilized stream reaches enter reaches with unprotected banks. And freshly quarried, angular rock riprap can have high rates of chemical weathering that release undesirable minerals to the water.[18]

A river's constant adjustments to changing water and sediment input imply that the river may not behave predictably. Failure to regard each river as unique can result in the blanket application of a particular rehabilitation technique that is not always appropriate. Several national forest units in the western United States spent hundreds of thousands of dollars each during the 1980s installing log-drop structures to create fish habitat along rivers flowing through alpine meadows. The

Failed rehabilitation attempt along Deep Run Creek, Maryland. Before rehabilitation, this stable, meandering stream had pools and riffles and well-developed riverside forest. An assessment of the stream during a period of low flow mistakenly interpreted the exposed riffle gravels as a sign of problems with sediment accumulation. The creek was recontoured with earth-moving equipment into a different channel form, and large logs were buried in the streambanks in an attempt to anchor this form. A flood after rehabilitation caused extensive streambank erosion, and the channel is now shallow and unshaded by riverside vegetation, with few pools. In this view, the formerly buried logs now protrude several feet out from the streambank into the channel.

structures consisted of large logs placed across stream channels and anchored into each bank. Large logs do not naturally fall into these streams, which meander across broad meadows. A decade after installation, the logs were undermined by streambed erosion or left in abandoned channels as the streams meander away from the structure. As Luna Leopold wrote in 2001, "Rivers do not construct drop structures. Rivers construct and maintain, by process of erosion and deposition, channels of particular characteristics . . . scaled to the size of the drainage basin and the nature of the rocks of the area."[19]

The idea of using drop structures to rehabilitate downcutting rivers dates back to the 1890s. The newly formed Soil Erosion Service was searching for means to control deep gullies forming on agricultural lands in many parts of the United States. Many of the engineers design-

View of a log-drop structure installed in an alpine stream in the Bighorn National Forest, Wyoming. The streambank has eroded around the log, which no longer forms a drop. (Courtesy of David Cooper.)

ing these structures believed that the structure would cause sediment accumulation and gully filling upstream from the structure all the way to the drainage divide. This initial hope was dashed as it became clear that each structure decreased the channel slope for only a short distance upstream and that the structure created an unnatural anomaly that would be removed sooner or later by undercutting or lateral erosion.[20]

Another problematic perception is the idea that a river can be restored to its initial condition simply by bulldozing a new channel form or stabilizing the streambanks. A river changes its form in response to changes in the water and sediment entering the channel reach from upstream or from adjacent hillslopes. It is futile to try to restore the altered channel to match some unimpacted nearby stream, or original condition, if the altered water and sediment supply that initiated channel change are still present. Local reaches of the Carmel River in California were widened from one hundred feet to nearly seven hundred feet between 1978 and 1980, and the river changed from meandering to braided. The regional water table had been lowered by groundwater pumping to the point that willows and other riverside plants died, weakening the streambanks and allowing massive bank erosion. Attempts to narrow and stabilize the river must rely on irrigation for the newly planted riverside vegetation because the original condition of the river can no longer be self-sustaining.[21]

The Carmel River exemplifies how a segment of river corridor is connected to all of the surrounding landscape, including the subsurface. Studies during the late 1980s and 1990s demonstrated the importance of the hyporheic zone, the shallow groundwater beneath a river channel. Many rivers have extensive exchanges between groundwater and surface water in upwelling and downwelling sites within the channel and across the floodplain. These sites exert an important control on water temperature and chemistry, and thus on biological productivity. If river restoration does not address groundwater tables and exchanges between the river and groundwater, full biological productivity and ecosystem diversity may not be restored to the river.[22]

Those trying to engineer rivers would do well to remember the principle guiding physicians: first, do no harm. Any successful rehabilitation attempt must be designed with close attention to a series of questions:

- Toward what condition are you trying to rehabilitate: conditions that existed before all human disturbance, or some subsequent state?
- What are the goals of rehabilitation: channel stability, flood conveyance, or habitat enhancement?
- How have the controlling factors of water and sediment yield to the channel changed?
- How will the proposed rehabilitation measures likely affect other aspects of river function beyond those explicit in the rehabilitation design?
- How will the rehabilitation design likely perform over time?
- Must other factors beyond physical characteristics, such as water quality, riverside vegetation, or exotic species, be addressed for successful rehabilitation?
- Is it possible to restore the channel to a predisturbance or highly desirable state?[23]

The most successful rehabilitation projects seem to be those that are least manipulative. Given sufficient time and space to migrate laterally, recontour its banks, and develop a riverside forest, an eroding river will eventually reach a new stable form and stop rapidly eroding its bed and banks. Provided with some approximation of natural flow patterns, and some sediment from upstream, and protected from excess sediment coming from adjacent lands, the river will likely stabilize in a form approximating its predisturbance condition. The problem, to paraphrase Robert Frost, is that we are often too anxious for rivers. We are impatient and invasive in our designs, recontouring the river into a "natural" form with riprap or anchored logs, carefully bulldozing a meandering path that mimics a sine-generated curve, and then expecting the river to remain within that path. Meandering rivers migrate laterally along the outside of each meander bend. To expect a carefully designed meander to remain unchanged through time is to ignore the fundamental processes inherent in rivers. And physical and biological recovery can be slow following rehabilitation attempts. Studies of channel recovery in abandoned mine lands that do not also have acid-mine drainage indicate that at least two decades may be required.[24]

Successful rehabilitation strategies also work with the river, rather than against it. As discussed in the previous chapter, experience with downcut channels in northern Mississippi indicates that grade-control structures placed in a rapidly downcutting river will be undermined by erosion, whereas structures placed in a river that is beginning to stabilize will enhance this process.[25]

Geographer Will Graf has proposed that we give more attention to the concept of physical integrity when planning and evaluating river rehabilitation. Rivers possess physical integrity when river processes and landforms are actively connected to each other under present flow conditions. In other words, the river and the adjacent floodplains constantly adjust to one another in a river with physical integrity, rather than being artificially fixed or isolated from one another by cement-covered streambed and banks or constructed levees. The river-floodplain adjustments of a rehabilitated system would occur within limits defined by societal values. If we restore the diversity of flow, channel form, and bed material to a river, then these diverse habitats will help reestablish its biological integrity. The presence and diversity of river invertebrates, for example, depend on water quality, streambed sediment, and flow velocity and depth, as well as sources of food. Fish are governed by water quality, cover, presence of food, and presence of habitat for spawning, egg incubation, and larval fish growth. Physical integrity alone is not enough to restore biological integrity if persistent contaminants or introduced species are present, or if a food resource or essential symbiotic species has been eliminated. Physical integrity is only one of many components necessary for restoring biological integrity.[26]

River rehabilitation is challenging, and there are many examples of poorly conceived rehabilitation projects that have failed. This does not mean that river rehabilitation should not be attempted. We are not yet especially skilled at undoing the damage our past activities have inflicted on rivers, but it is imperative that we learn these skills by practicing them. We need to improve communications among those doing river rehabilitation so that mistakes are not repeated indefinitely, and we need studies evaluating the long-term success or failure of past rehabilitation attempts. As Luna Leopold wrote in 2001: "There are a lot of people harming rivers. There are also people who are improving them. But we do not know who is doing what."[27]

One of the more ambitious rehabilitation projects in the United States, and one receiving much scrutiny, involves the Kissimmee River of Florida. The scientific, political, and socioeconomic skills developed in the course of this project will undoubtedly prove vital in future projects.

Bringing Back a River of Grass

Florida is flat. The highest point in the state is only 325 feet above sea level. Along the course of the Kissimmee River in southern Florida, the adjacent "uplands" are only 6 to 10 feet higher than the floodplain. The Kissimmee River once meandered in broad arcs across this floodplain, dropping 6 to 9 feet for every 100,000 feet traveled along its 100-mile-long drainage basin. Native Americans named the Kissimmee for its "winding waters."[28]

From the air, the Kissimmee drainage basin looks as though someone had thrown a handful of pebbles through a layer of brilliant green pondweed, leaving many little clear holes to the water beneath. These holes are karst lakes, for the "pond" beneath the green of the basin is a vast layer of carbonate rocks honeycombed with water-filled caves and sinkholes. As the continental ice sheets melted about 128,000 years ago, water released into the oceans raised sea levels. Southern Florida became a tropical lagoon in which corals grew, and the carbonate shells of tiny marine creatures slowly settled to the seafloor. During the next Ice Age, receding seas deposited coastal sand ridges across the carbonates.

Aerial view of a portion of the Kissimmee River in Florida before channelization in 1961. Note the sinuosity of the river and the multitude of secondary channels. (Courtesy of the South Florida Water Management District.)

These sand ridges subsequently formed highlands once the whole area was exposed above sea level. The landscape assumed its present form since sea level stabilized about 5,000 years ago.[29]

Once the carbonates of southern Florida lost the protection of the ocean, they slowly began to dissolve. Most rainwater is slightly acidic, and it reacts with carbonate rocks, dissolving the calcium from the rock and carrying the calcium away in solution, leaving sinkholes. Water fills the sinkholes to create lakes and spills across the lowlands to form rivers and adjacent wetlands. Before European Americans intervened, the different components of this giant sponge were subtly but intricately connected. Rain that fell on the interior was carefully used, soaked up, and filtered through a huge landscape and passed on from organism to organism, nourishing the rivers, lakes, prairies, and wetlands. Black, organic-rich water drained slowly into Lake Kissimmee in the middle of Florida and then down into the Kissimmee River. During wet periods the waters flowed south in a broad band, inundating a floodplain up to three miles wide. Eventually, the waters coalesced into Lake Okeechobee ("big water" in Seminole) and then spilled out once more to filter down through the Everglades in broad sheets. Rains during the

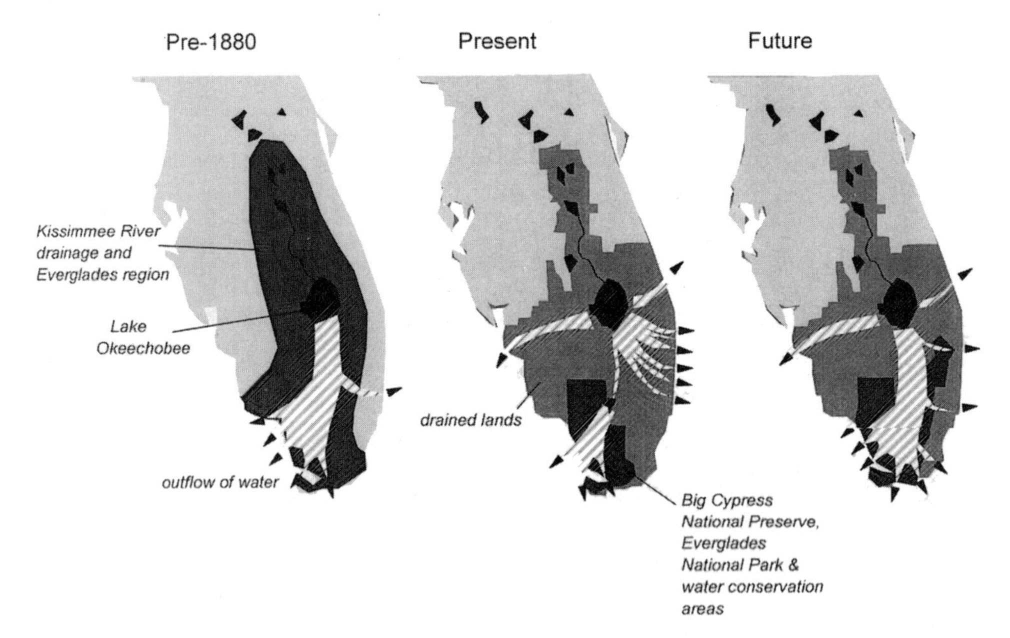

Water flow paths for historical and contemporary conditions and the future after restoration in southern Florida. Together, the Kissimmee-Okeechobee-Everglades system covers 10,890 square miles—310 miles north to south and 62 miles east to west. Southeast of Lake Okeechobee, three surface-water impoundments store 1,882,000 acre-feet of water, and two additional impoundments serve as groundwater recharge points. Between these impoundments and the lake lie 1,181 square miles of agricultural land with fifteen canals and twenty-five water-control structures. (After C. Chang, 2001, Go with the flow, *Audubon*, 103, 58–59.)

summer and autumn covered the floodplain for three to nine months each year, and periodically for the entire year. This flooding supported an annual invertebrate production on the floodplain up to one hundred times greater than the production in the stream channel. As water levels declined each year, the invertebrates were carried into the river, where they served as food for many other animals. The life cycles of water birds and fish were also linked to flooding. Wood storks ate fish concentrated at water holes during the winter, and snail kites fed on apple snails whose egg laying was tied to seasonal water fluctuations. Fish used the floodplain habitat for spawning and nursery areas. Years with a smooth increase in water level, and with large floods that lasted a long time, were good years for the thirty-five species of fish in the region.[30]

Over most of the United States, surface water is confined. It flows within defined river channels that occupy only a small percentage of the

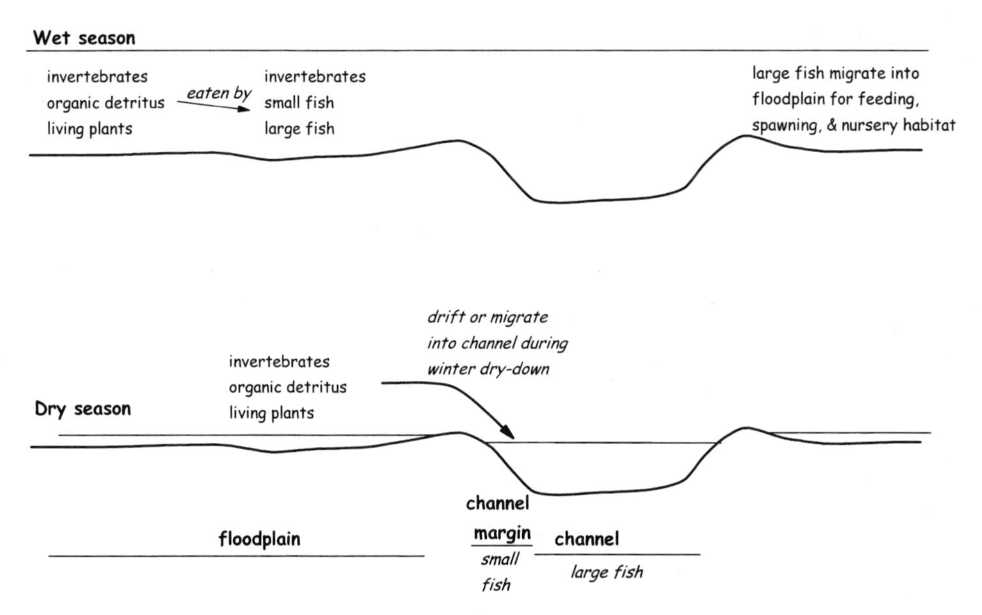

Schematic cross-sectional view of relationships between the channel and floodplain along the Kissimmee River. Invertebrates, detritus, and plants originate in the floodplain and are consumed there throughout the year by other invertebrates and by small fish such as mosquito fish or killifish. During the wet season they are consumed by large fish such as carp, tilapia, gar, bowfin, bass, sunfish, and chain pickerel that migrate from the channel onto the floodplain. As water recedes from the floodplain during winter, some of the remaining invertebrates and detritus moves into the channel and provides food for the fish.

total landmass, or it ponds in clearly defined lakes. In the Kissimmee River drainage basin this relation is reversed, and waters dominate the land. Despite the flatness, islands are present in the form of hardwood hammocks. The southwest coast of Florida is called the Ten Thousand Islands, but this might be the name for all of southern Florida. Where the underlying bedrock is even a few inches higher, saw grass marshes once formed. The decaying plant material built low mounds that were then colonized by shrubby plants such as wax myrtle and button bush. As peat continued to accumulate, willows, pond apples, and gumbo-limbo trees became established in the centers of these hammocks, which grew to be up to two miles long. To this haven came wading birds to nest; alligators and turtles to lay their eggs; and deer, snakes, and lizards seeking refuge during periods of rising water. Two to three times more plant and animal species used the tree islands than used the surrounding marsh.[31]

The surrounding marsh was hardly underutilized, however. The land of water was biologically rich. As late as 1934, a quarter-million white ibis nested in a single colony in southern Florida. They represent all the others—wood stork, great blue heron, roseate spoonbill, dozens of species of neotropical migrants, crested caracara, snail kite, ibis, great egret, thirty-eight species of wading birds and waterfowl—who live here or use the area as a crucial stop on their annual migrations. And in the waters and islands below the birds are the endangered Florida panther, gopher tortoise, eastern diamondback rattlesnake, bluetail mole skink, crocodile, alligator, Florida gar, and many other species—a plethora of life relying on the dynamics of the rivers of grass.[32]

People were inevitably drawn to this rich landscape. Hunter-gatherers occupied the area as early as twelve thousand years ago. As the large game animals disappeared, agricultural villages developed about four thousand years ago. When Spaniard Ponce de Léon christened La Florida for the land of flowers in 1513, he found the Cayusa tribe in residence. The Cayusas were part of the estimated one hundred thousand Native Americans living in Florida at the time of Spanish contact. They built oyster-shell islands and dug canals through the wetlands, beginning the alteration of this water-logged world to suit human purposes. The Cayusas were strong, well-made people, but they had little resistance to the microbes fostered by generations of high human population density in Eurasia.[33]

The Cayusa population was drastically reduced by epidemics of diseases introduced by the Europeans, exacerbated by war and slavery. Europeans quickly sought to develop their own civilizations in the newly depopulated lands. The Spanish began to plant sugarcane at their settlement of St. Augustine in 1572, and British and other European explorers arrived. The expanding American colonies to the north displaced Native Americans who became known as Seminoles when they moved into Florida in the late 1700s, about a century before the American colonists arrived. The Seminoles called the region Pay-haio-kee, for "grassy water," and they developed their own agricultural society there. Pay-haio-kee was supplanted by the name Everglades, coined in the 1820s from the Old English "glyde" or "glaed," signifying an opening in the forest. Despite all the preceding human activity, John James Audubon could still perceive southern Florida as largely wilderness as late as his 1832 visit.[34]

During the second of the three wars between the United States and

View of the Kissimmee River near Fort Basinger Station in 1919 by John K. Small, showing channel and floodplain filled with water. (Courtesy of the Florida Department of State, Division of Library and Information Service.)

the Seminoles in 1835, U.S. soldiers began to leave written descriptions of southern Florida. The region remained largely unknown to European Americans, despite the Spaniards and Audubon. Ten years later Florida became the twenty-seventh state, its population recorded as seventy thousand. When Congress passed the Swamplands Act in 1850, twenty million acres were transferred to the state for drainage and reclamation. Little was done until a Philadelphia millionaire bought four million acres in 1881, the first of many subsequent real-estate transactions in Florida. Within ten years, fifty thousand acres had been drained and eleven miles of canals built, starting the massive landscape transformation that occurred during the next century.[35]

By 1900, the population of the lower east coast of Florida grew to nearly twenty-three thousand. Some of these people hunted feathers for the millinery trade. These hunters were anything but selective. Amid

Early settlers in the Kissimmee River region, circa 1890s. (Courtesy of the Florida Department of State, Division of Library and Information Service.)

fears that the Everglades rookeries would be decimated, the Audubon Society facilitated a law in 1901 prohibiting bird hunting in much of the region. Four years later the first Florida game warden was appointed. The sparsely settled, physically intimidating lands of the interior and the southwestern coast had a state of lawlessness similar to that of the fabled Old West. The game warden was promptly shot to death.[36]

After World War II, human population and infrastructure began to increase more rapidly in the Kissimmee River drainage basin. A major hurricane in 1947 resulted in extensive flooding and property damage, and the Army Corps of Engineers stepped in. Congress authorized the Kissimmee River Flood Control Project in 1954. Between 1962 and 1971, the corps dredged a trapezoidal canal from Lake Kissimmee to Lake Okeechobee. The canal is about ten times the size of the natural channel, with six water-control structures that regulate water levels and flow, creating a reservoir upstream of each structure. In most areas of the world, roads go downhill into a valley to cross a river. In southern Florida, the rivers are walled within levees that sit above the surrounding lands, and roads rise up to cross a river. Similarly, Lake Okeechobee is "moated"

Aerial view of the State Road 70 bridge across the Kissimmee River floodplain during the 1948 flood. (Courtesy of the South Florida Water Management District.)

and gated, surrounded by a high levee with occasional gates so that the lake is perched above its surroundings.[37]

The Kissimmee River Flood Control Project did alleviate flooding, but it also largely destroyed the floodplain ecosystem. Dredge material from the canal buried sixty-nine hundred acres of floodplain wetlands. Another thirty-five thousand acres were altered by the loss of seasonal flooding. Remaining segments of the natural channel carried little or no flow. As several inches of organic muck accumulated over the sandy bottoms of these channel segments, the stagnant water became anoxic, with little oxygen available to aquatic plants and animals. Levels of phosphorus entering Lake Okeechobee increased from one hundred to five hundred tons per year as dairy, citrus, ranching, and sugarcane operations spread through the former wetlands. The excess nutrients created algal blooms that depleted oxygen levels in the lake waters, destroying the abundance of autumn insects that had fueled migratory birds on their way.[38]

Channelization destroyed or degraded most of the fish and wildlife habitat provided by the river and its floodplain wetlands. The maintenance of stable water levels within the remaining wetlands eliminated

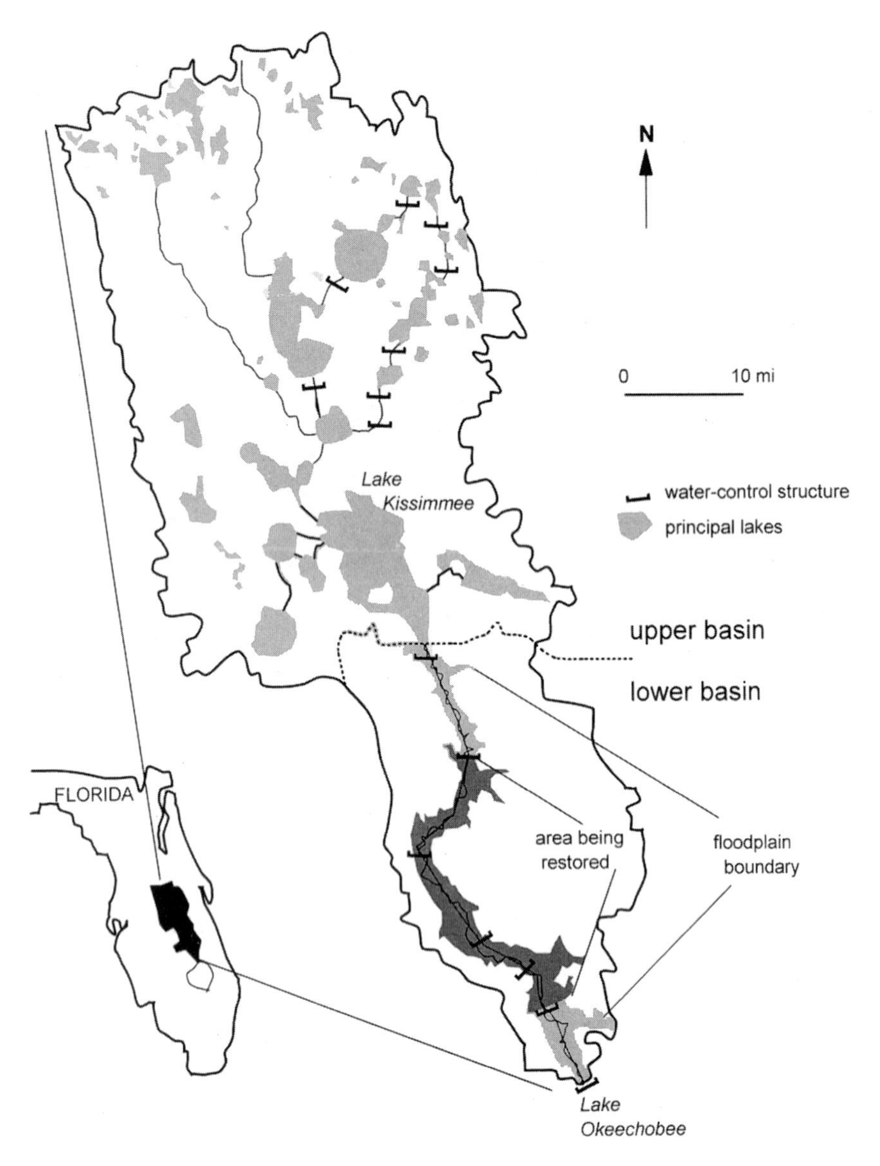

Kissimmee River drainage basin, with locations of major water-control structures and extent of river and floodplain restoration project. (After A. G. Warne, L. A. Toth, and W. A. White, 2000, Drainage-basin-scale geomorphic analysis to determine reference conditions for ecologic restoration—Kissimmee River, Florida, *Geological Society of America Bulletin,* 112, 884–99, Figure 2.)

View of spillway, lock, and dam S-65E along the Kissimmee River, 1964. Built at a cost of more than $1 million, this is one of six lock, dam, and spillway structures along the ninety-seven-mile-long Kissimmee Waterway. All locks measure thirty feet by ninety feet and are designed to pass vessels of five-and-one-half-feet draft year-round. (Courtesy of the Florida Department of State, Division of Library and Information Service.)

the diversity and spatial heterogeneity once provided by a mosaic of vegetation types. The pulselike flows of the channelized river had high and low flow periods out of phase with typical seasonal flow patterns present before channelization. As a result, wading birds used the floodplain much less. Numbers of wintering waterfowl dropped by 92 percent. More than five billion small forage fish were lost, several indigenous fish species were destroyed, and game fish such as largemouth bass declined significantly.[39]

Almost as soon as channelization was complete in 1971, public outcry galvanized political resolve to restore the Kissimmee ecosystem. The Florida legislature passed the Kissimmee River Restoration Act in 1976. By 1983, Florida governor Bob Graham had created a coordinating council to oversee the restoration as part of his Save Our Everglades initiative. A massive rehabilitation project was designed with the goal of restoring

View of the construction of the S-65 gates along the Kissimmee River, 1965. (Courtesy of the South Florida Water Management District.)

The channelized Kissimmee River just upstream from Lake Okeechobee. Here the tea-brown water flows in a broad, arrow-straight channel.

Sign at bridge crossing along the channelized Kissimmee River.

the ecological integrity of the river ecosystem. Ecological integrity of the Kissimmee River is to be judged by five factors. These include energy source, which depends on inputs of organic matter. In other words, is there enough mass of plant material to produce the nutrient-rich muck on which all other life depends? A second factor is water quality, judged in terms of temperature, turbidity, dissolved oxygen, and other characteristics. Is the water warm and clear and oxygenated enough to support insects, fish, turtles, and alligators? Habitat quality, as judged by streambed composition, flow depth, flow velocity, and diversity, forms a third factor. Hydrology, measured as flow and variability of flow through time, is the fourth factor. Finally, biological interactions, including competition, predation, disease, and parasitism, must be restored.[40]

Project scientists established flow criteria that they believed would reestablish "an ecosystem . . . capable of supporting and maintaining a balanced, integrated, adaptive community of organisms having a species composition, diversity and functional organization comparable to the natural habitat of the region." The criteria necessary to restore these habitats include continuous flow from July to October. Highest annual flow is to occur during September to November, and lowest flows during March to May. In addition, the criteria specify a wide range of variability in flow between years. These flows should maintain favorable dissolved oxygen levels during summer and autumn. These conditions should also provide nondisruptive flow for fish during spring reproduction, and restore habitat heterogeneity across both time and space. In practice, these criteria became very precise, with specific requirements for flow velocity, depth, and timing.[41]

In order to meet these criteria, program managers evaluated three plans, each of which proposed a different method to divert water from the canal back to the remnants of natural channels. The first plan involved placing ten weirs next to the ten most suitable channel remnants to be revitalized in the Kissimmee River. Instead of weirs, the second plan proposed using long earth plugs to divert water from the canal to the channel remnants. The third plan focused on backfilling a long, continuous reach of the canal, thus forcing flow into the natural channel remnants. Managers eventually decided to remove two of the existing six water-control structures, backfill twenty-two miles of canal, reexcavate portions of the river channel that were destroyed, and purchase about seventy thousand acres of floodplain through the State of Florida, all at an estimated cost ranging from $280 million to $422 million. This is one component of the $7.8 billion Comprehensive Everglades Restoration Plan, a massive experiment to determine whether we can in fact undo our negative impacts. Money is not everything, but it is a crucial component of the restoration process. A National Research Council report issued in 2003 noted that inadequate funding was hampering Everglades restoration efforts.[42]

Managers installed three weirs across a portion of the canal below Lake Kissimmee between 1984 and 1989 as a demonstration project. The weirs simulated the effects of dechannelization by diverting flow into remnant river channels and floodplains. The demonstration project had some encouraging successes. The diverted flows increased dissolved oxygen levels in the river water. The flows carried downstream the fine sediments that had accumulated in the river since channelization and restored the sandy streambed. The flows also recontoured the uniformly flat, shallow channel into a channel with pools and sandy riffles. As the physical integrity of the river was restored, biological integrity also began to recover. Bottom-dwelling invertebrates became more numerous and diverse. Game fish became more abundant. But the demonstration project also indicated the limitations to river rehabilitation. More

Sheet pile weir used to divert water from the canal back into a remnant of the natural channel (to the left) as part of the demonstration project on the Kissimmee River. (Courtesy of the South Florida Water Management District.)

complete restoration of the river's biological integrity requires the reestablishment of historical flow patterns. As cautiously summarized by the editors of a special scientific volume on the restoration effort, "with all our expertise in ecosystem restoration, it is widely recognized that it is unlikely that the full spectrum of structural and functional attributes of the system can be restored to the levels existent prior to the disturbance."[43] All the king's horses and all the king's men couldn't put Humpty together again . . .

Reestablishment of historical flow patterns will not come easily, for many people now rely on southern Florida's waters. The 1900 population of approximately 23,000 grew to more than 228,000 by 1930. By 2001, the population of southern Florida reached 6 million, with projections of 12 million by 2050. The 15,000 acres of sugarcane being grown in the northern Everglades in 1933 expanded to 450,000 acres by 1990, and half of the original wetlands were converted to sugarcane, other types of farming, and housing. Tourism and seasonal residence have boomed. In winter, the flow of people from the eastern and midwestern United States south to Florida is as impressive a phenomenon, in

Aerial view of meanders isolated from the main channel of the Kissimmee River by channelization. This meander is adjacent to one of the demonstration project weirs and now has water once more. (Courtesy of the South Florida Water Management District.)

its own way, as the concentration of waters across nearly two-thirds of the country into the southward flow of the Mississippi River.[44]

As humans and infrastructure increased, wildlife decreased. Some of this decrease resulted from deliberate human actions; some was inadvertent. During the nineteenth century, Victorian ladies and gentlemen kept curiosity cabinets filled with shells or feathers or rocks. Snail collectors supplying these cabinets realized that each hardwood hammock in the Everglades harbored a distinctive species of snail with a unique shell. Some collectors took all the snails they could find and then burned the hammock so that no one else could ever duplicate their collection. Other animals fell to the guns of plume hunters, or the axes of lumber workers. The ivory-billed woodpecker was last sighted in southern Florida during 1917. The Florida panther, snail kite, and Cape Sable seaside sparrow were listed as endangered species in 1967. They were joined by the American crocodile in 1975 and the wood stork in 1984.[45]

The political response to these diverging trends of human and wildlife populations began with the 1934 reservation of more than two million acres in Everglades National Park. The park was enlarged in 1947, the year that Marjory Stoneman Douglas's wonderful book *The Everglades: River of Grass* brought the region to national attention.[46]

At the start of the twenty-first century, Everglades National Park is like a small water hole during a drought—concentrating life, with less and less hope of preserving it. There was always a concentration of life around actual water holes during the dry season. Historically, these were sinkholes or alligator holes. Now they are as likely to be quarries left from roads built on long mounds of crushed rock to protect the roads from flooding in this flat, watery land. But as the seasonal flow of water steadily decreases, the concentration of life around water holes takes on the urgency of irrevocable change. Lightning-generated fires historically burnt away the undergrowth in the pinelands during the dry season, for example, keeping the hardwoods from replacing the pines. High water tables kept the soils moist and the fires at low intensity. As water tables decline, the fires burn hot enough to threaten the pines. The fires even burn the peatlike soil itself, reducing the ability of any plants to germinate after the fire. Abundant water is crucial to the functioning of this ecosystem. Restricting the extent and volume of water is like restricting the circulation of blood to a limb; the limb ceases to function.

It becomes difficult to escape the image of Everglades National Park as an embattled and besieged remnant. Every interpretive display at the park emphasizes declining wildlife populations because of flow regulation and toxic contaminants. High levels of mercury are present throughout the food web, from soils to panthers. An endangered Florida panther, of which fewer than thirty may still be alive in all of Florida, was found dead in the park. Its body contained mercury levels that would be toxic to humans.[47]

Interpretive displays at the park also feature the different perspectives of participants in the great South Florida water controversy: a homeowner who wants to water her lawn and to reduce the taxes she pays to subsidize water engineering for farmers; a vacationing sports enthusiast disgruntled at the decline of the once-legendary Lake Okeechobee fishery; a commercial fisher working in the Gulf of Mexico, facing loss of livelihood as Gulf productivity declines; an environmentalist seeking to restore seasonal flow fluctuations and clean water to southern Florida; a farmer warning that citrus and other fruits and vegetables cannot be grown without the water engineering that the government invited his father to come use; and a park ranger citing the declines in wildlife populations in the park. Nobody wants to destroy the Everglades, or the Kissimmee, or Lake Okeechobee. Yet in this region that once overflowed with an abundance of water—from the sky, on the land—we have created water shortages, and we argue bitterly over who gets the remaining water.

Amidst these arguments, citizens form groups called Save Our Everglades or Everglades for Everybody to promote their viewpoints. Meanwhile, the human population of southern Florida continues to grow, and attempts to restore the Everglades and the Kissimmee River move slowly forward. Asked to develop an evaluation program for the Kissimmee Restoration Project, a scientific advisory panel recommended five phases of evaluation. First, establish reference conditions that define realistic expectations for restoration. Next, establish baseline conditions that define the current state of the ecosystem. Third, assess the effects of construction in order to minimize them. Fourth, assess the short- and long-term responses of ecosystems to restoration. And finally, implement adaptive management to provide continuous fine-tuning of restoration efforts. These procedures and the manner of their implementation will likely form a blueprint for any future large-scale attempts at river restoration.[48]

As scientists, environmental activists, and politicians prepare to implement the Kissimmee River restoration, studies are underway to improve our understanding of the life histories of the various species at risk. Among these is the American alligator, which has already survived one brush with extinction.

Gravel Worms

For some one hundred million years, reptiles of the order Crocodilia have divided their time between the land and the water. The name crocodilia is derived from the Greek *krokodilos,* signifying "gravel worm." Alligators are included this order, the name alligator having come from the Latin *lacertus,* for "upper arm." The Romans used this word to describe lizards, which they associated with the shape and size of an arm. Today, only two species of alligators remain, an endangered or perhaps extinct species in China, and *Alligator mississippiensis* in the southeastern United States.[49]

An alligator basking in the winter sun along a water-filled borrow pit created during road construction in Everglades National Park.

The alligator is a large animal that dominates both its watery landscape and the imagination of people who venture into that landscape. The sight of an elongated head and ridged, leathery back floating in the dark water triggers an instinctive revulsion and fear in many people. The big reptiles can disappear beneath the water, leaving a disconcerting sense of being watched by hidden eyes. A large alligator threatened by humans too close to the alligator's nest is quite capable of dragging them underwater, drowning them, and then devouring them in pieces. Yet on sunny winter days, the alligators basking at evenly spaced intervals along the banks of a pond are so unmoving that they might be museum specimens. Baby alligators only a few inches long lie in disordered heaps around each mother's head. There is such an absence of menace that it is tempting to reach out and touch a leathery black tail. Then an eye opens. At night, the water comes alive under a flashlight as the eyeshine from dozens of alligators reflects the light like floating lanterns.

Male alligators average eleven feet in length and can reach a weight of nearly half a ton. The females are hardly more diminutive, at an average length of eight feet. Both sexes have a body built for power: thick limbs, broad head, and half of its length in a strong tail used to propel the alligator through the water. Despite these abilities, alligators mostly live on prey smaller than humans. Fish, turtles, snakes, snails, birds, and small mammals such as raccoons and nutria form most of their prey. Biologists have described alligators as eating "everything that moves and some things that do not." The alligators are most agile in the water, so it is from water that they snap up the small prey and swallow it whole, or lunge up and drag the larger prey back into the water with them.[50]

Alligators prefer fresh to brackish water, and they live in the river swamps, lakes, bayous, and marshes of the Gulf and lower Atlantic coastal plains from Texas to North Carolina. In these environments the alligators dig a den at the edge of a river or lake, with an underwater entrance that leads as much as twenty feet back to an underground hollow partially above the water level. The Everglades provide a harsh environment for alligators, as evidenced by the facts that alligators there have smaller average clutch sizes and reach maturity at later ages. Temperatures remain high year-round in the Everglades, keeping the cold-blooded alligators at higher metabolic levels, but seasonal food shortages make it difficult to fuel those metabolisms. One response is to dig gator holes. Adults create and maintain these holes over a period of years. Using mouth and claws to uproot vegetation, the alligator then shoves with the body and slashes with the tail to wallow out a depression that will remain filled with cool water during the dry season and periods of drought. Others come to these alligator oases, too. The gator holes provide water for fish, insects, crustaceans, snakes, turtles, and birds. The big reptiles thus shape both the plant and animal communities around them by altering habitat with their holes and trails, and altering species composition through predation.[51]

Alligators also create mounds in the landscape. With the coming of spring, the big males begin to roar to attract mates and warn off rivals. Although they do not have vocal cords, they make plenty of noise by sucking air into their lungs and blowing it out in deep, intermittent bellows. Courtship and breeding continue from April into mid-June. The males then remain in deep water while the females move toward land to build their incubation mounds. After laying twenty to fifty eggs, the female piles vegetation and mud over them in a mound nearly two feet high and six feet wide. The heat of the decaying vegetation warms the

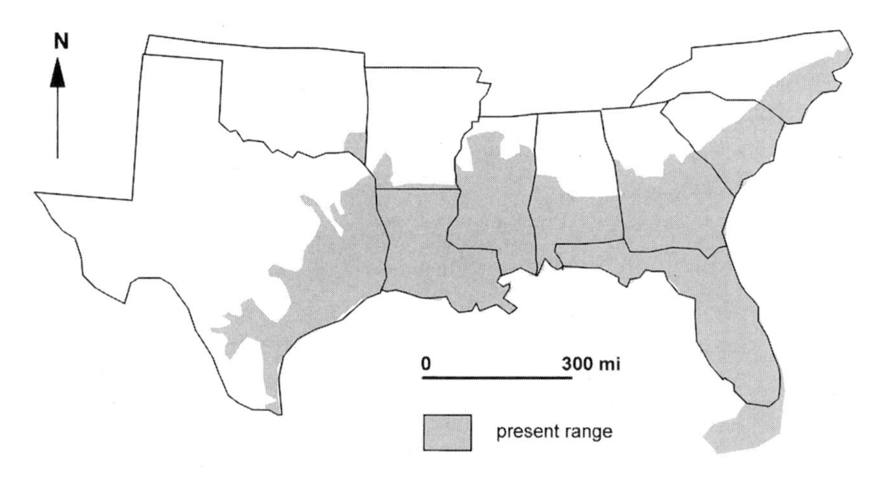

Range of the American alligator in the United States. (After F. J. Mazzotti and L. A. Brandt, 1994, Ecology of the American alligator in a seasonally fluctuating environment, in S. M. Davis and J. C. Ogden, eds., *Everglades: The ecosystem and its restoration,* St. Lucie Press, Delray Beach, Fla., Figure 20.1.)

eggs while the mother stands on guard to protect the nest during the sixty-five days of incubation. Temperature is crucial during this period, for alligators have no sex chromosomes. The development of their gonads is controlled by the incubation temperature. Males result from higher temperatures of 90.5 to 93 degrees Fahrenheit, and females from lower temperatures of 82 to 86 degrees Fahrenheit.[52]

When they are ready to hatch, the little alligators emit high-pitched croaks that bring their mother to the rescue. She digs out the babies, and the six- to eight-inch-long nestlings head immediately for the water. Although they have needlelike teeth, they live for several days on yolk masses within their bellies. As they grow, they eat insects, crustaceans, snails, and small fish. The mother protects the young ones for several months, and they live together in small groups, but they are vulnerable. About 80 percent of the new alligators are eaten by birds, raccoons, larger fish, snakes, bobcats, otters, and even larger alligators. The alligators that survive grow about one foot per year, reaching breeding maturity at eight to thirteen years of age. They may live more than thirty years, if something doesn't intervene. For the past few decades, that something has usually been a human.[53]

Young alligators are brightly patterned in black and yellow. As they mature, their skin fades to dark gray or black. But this duller hide is still very attractive to humans, who use it to make leather. Alligator meat is also in demand. By 1960, trapping and market hunting reduced alligator populations in the United States to a dangerously low level. The animal was listed as an endangered species in 1967, under a law that preceded the 1973 Endangered Species Act. The alligators responded, and in 1987 the Fish and Wildlife Service pronounced the animal fully recovered and removed it from the list.[54]

The alligator may not be free and clear, however. Habitat loss is now the chief threat. An adult needs at least thirteen acres of habitat, and both males and females are highly territorial. Males occupy summer home ranges with an average minimum size of nearly twenty-two hundred acres, whereas females need an average of only twenty-one acres.[55]

Habitat quality is also very important to alligator survival. Patterns of courtship, mating, nesting, and other habitat use all depend on wetland water levels. Drought conditions, whether natural or induced by human regulation of river flow, can expose females guarding nests to metabolically stressful heat. Flooding can drown the nest. Numerous studies indicate that alligators in the United States are stressed by the drainage of wetlands and the alteration of natural flow fluctuations in remaining wetlands.[56]

More insidious are the chemicals released into the environment as a result of human activities. A 1998 study of alligators living around Lake Apopka in central Florida found contaminant residues of DDE, methoxychlor, dieldrin, PCBs, and other chemicals in alligator eggs. High residue levels in eggs correlate with low egg viability and reduced hatchling success. DDE, dioxin, and other endocrine disrupters also cause alterations in the sexual characteristics of the hatchlings. Females hatch at male-producing temperatures. Male hatchlings have testosterone concentrations significantly lower than normal. The herbicide atrazine induces gonadal activity in male hatchlings that is characteristic of neither males nor females. By reducing the habitat in which alligators can live and reproduce, and reducing even their biological ability to reproduce, we may once again be placing the species in peril.[57]

Threats to alligator populations from chemical contaminants or overhunting indicate that restoring functional river ecosystems with a full complement of riparian and aquatic species requires more than just restoring physical processes along a river. A connected, fully functional river is influenced by the entire drainage basin. Restoring processes across the drainage basin is much more complex.

It is inevitable that we attempt to manage rivers. But the hubris of many of our present restoration efforts—in which we assume that we can build a river "from scratch" or impose an idealized form on a river—is analogous to our reckless and ill-advised forays into the chemical control of pests in its impatience, immature thought, and adverse effects. To borrow an old adage, expecting river rehabilitation efforts to be careful, thoughtful, and individually tailored is not blind opposition to progress, but opposition to blind progress.

Our restoration efforts in the Kissimmee River basin will require an enormous investment of time and money. We will undoubtedly learn much from these efforts, but ecologists warn that the Kissimmee example cannot necessarily be used to judge our ability to restore other river ecosystems. The Kissimmee River is in some ways a biologically simpler system to restore. First, only two of thirty-four exclusively freshwater fish species present in the river before channelization were not collected during a 1986–91 survey. This indicates that the original fish community, although depleted in numbers, is still present at some level and presumably could recover if physical habitat were restored. The two missing species, the coastal shiner and the blackbanded darter, are abundant in nearby habitats. Second, nonnative fish are not currently abundant in the Kissimmee River and thus are not likely to impede efforts to restore native fish species. In contrast, channelized river systems that originally had high species diversity, and endemic species and unique races, are less feasible to restore to prechannelization conditions than is the Kissimmee River. We need to keep trying to restore and rehabilitate rivers, but our efforts must reflect knowledge, patience, and a willingness to learn from past mistakes. Our rivers deserve nothing less.[58]

Chapter 7

Thinking in Terms of Rivers

Endangered U.S. Rivers

Rivers of the six American regions described earlier in this book share many of the same impacts, although to differing degrees. Each region also has impacts that are particularly widespread and intense in that region. The following regional summaries include broad generalizations, and each region holds at least one relatively unimpacted river.

Rivers of the Northeast and East-Central region have the longest history of intensive European American land use in the United States. Widespread deforestation and cropping, as well as construction of small dams, commercial fishing, and industrialization and urbanization, have affected this region for more than two centuries. Flow volume remains close to natural levels along most of the rivers in the Northeast and East-Central region, but water pollution and habitat loss are ubiquitous. The primary sources of contamination are past and present industry and continuing urbanization. Small dams, such as Edwards Dam on the Kennebec River in Maine, are being removed. Many of these dams were built more than a century ago for a use that is no longer important to contemporary society. As these dams are removed and water quality improves, anadromous fish are once again gaining access to historical spawning grounds. However, two of the ten rivers listed by the organization American Rivers as being the nation's most endangered rivers for 2003 are within this region. The Ipswich River of Massachusetts is listed because groundwater pumping and excessive water consumption cause a portion of the forty-mile-long river to go dry each year. The Mattaponi River of Virginia made the list because a proposed water-supply reservoir threatens the ecological integrity of what is one of the most pristine coastal rivers on the eastern seaboard. These examples clearly illustrate that there is no room for complacency, despite the relative abundance of water in this region compared with the western United States, and despite the fact that some dams are being removed. An assessment of conservation status conducted by the World Wildlife Fund in 2000 listed rivers of this region as relatively stable in the north to critical in the south.

Rivers of the Lower Mississippi region were first heavily impacted by dramatically increased sediment yields from lands being cleared of natural vegetation for growing crops. River response to sedimentation prompted humans to undertake widespread channelization, construction of levees, clearing of logjams and naturally occurring wood, and destruction of riverside and floodplain habitats. During the twentieth century, the downstream-most portions of this region became notorious as the unofficial national dumping grounds for industrial and agricultural toxic wastes. Consequently, although the rivers in this region generally have abundant water, the water, sediments, and aquatic organisms are likely to be highly contaminated. The Big Sunflower River of Mississippi is on American Rivers's 2003 top ten list because of a massive flood-control project proposed by the Army Corps of Engineers. The so-called Yazoo Pumps will drain seven times more wetlands than private developers damage in a year nationwide, as well as dredging one hundred miles of the bed of this river that supports one of the world's most abundant native mussel beds and fifty-five species of fish. The Tallapoosa River of Alabama and Georgia made the list because proposed hydropower and water-supply dams would destroy some of the remaining healthy portions of a river where existing dams have essentially killed a forty-seven-mile stretch of channel by creating rapid and repeated alternations between dry riverbed and flood conditions. The World Wildlife Fund listed rivers of this region as mostly in critical conservation status.

Rivers of the Central region have been most altered by flow regulation and changes in sediment supply associated with dams, diversions, and channelization and by contamination from agricultural runoff. The central plains are arguably the least appreciated and noticed of America's natural landscapes, and most of the rivers of this region have attracted less attention than the scenic mountain and canyon rivers to the west. The central grasslands are also the most altered and endangered ecosystem in the United States. By 2003, 90 percent of the original 145 million acres of tallgrass prairie were replaced by other vegetation or

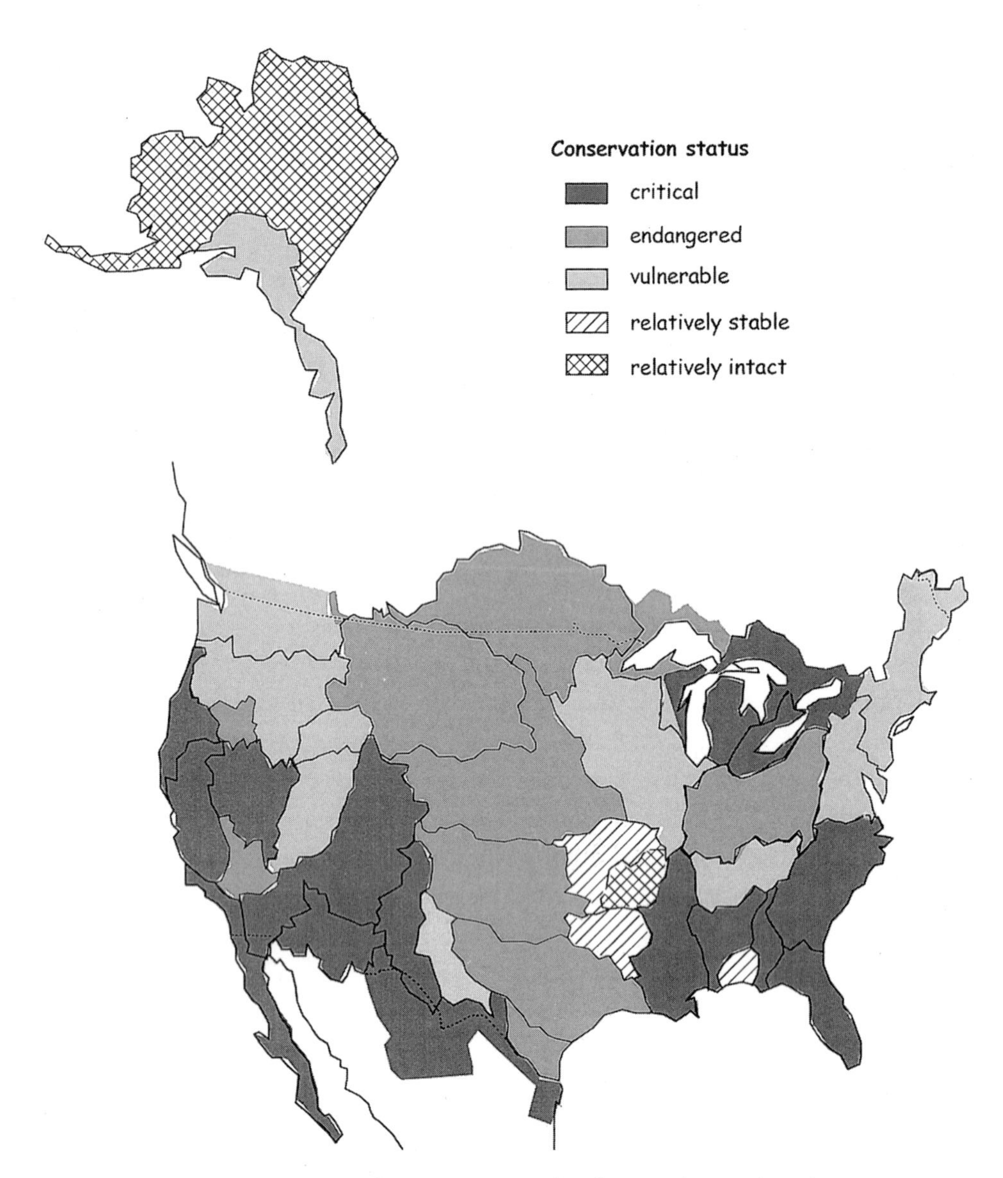

Conservation status of watersheds and regions within the United States, based on six factors: degree of land-cover alteration within the catchment, degradation of water quality, alteration of hydrologic integrity, degree of habitat fragmentation, effects of introduced species, and impacts of direct species exploitation (for example, fishing). (After R. A. Abell et al., 2000, *Freshwater ecoregions of North America: a conservation assessment,* Island Press, Washington, D.C., Figure 4.10.)

land use. The rivers of the grasslands reflect this history of alteration. The 2003 American Rivers list includes the Platte River of Wyoming, Colorado, and Nebraska. Continuing irrigation and water-supply development in the dry plains of the Platte River basin threaten to undermine existing agreements to protect in-stream flows and adjacent wetlands along the river that are crucial to tens of millions of migratory birds. The Trinity River of the eastern plains of Texas made the list because the Corps of Engineers proposes to undertake a flood-control project that includes constructing new levees, rerouting a portion of the river in Dallas, and replacing some of an 8,500-acre native floodplain forest with engineered floodplain. In the World Wildlife Fund assessment, rivers of the Central region range from relatively stable to vulnerable.

The big rivers of the Southwestern Canyon region were most heavily affected by the construction of numerous large dams and the removal of water for off-stream agricultural and municipal uses during the twentieth century. Today, this region has the nation's highest ratio of stored water relative to natural flow. Smaller streams in the region have been alternately cutting down and filling for more than a century, at least partly in response to grazing, groundwater withdrawal, and urbanization. Between altered flows and channel form and invasive species, the riverside and in-channel ecosystems of this region have been extensively modified. Contamination from agricultural and urban runoff is locally present. Human alteration of the Colorado River system has been nationally recognized for decades, thanks to the presence of the Grand Canyon and to books such as *A River No More* by Philip Fradkin and *Cadillac Desert* by Marc Reisner. Colorado's Gunnison River is on the 2003 list of most endangered rivers because the Department of the Interior under the Bush administration is reversing past efforts to protect in-stream flow within Black Canyon of the Gunnison National Park, opening the way for water transfers to the rapidly growing urban areas near Denver. The Rio Grande of Colorado, New Mexico, and Texas is on the list because of excessive diversion and overconsumption of water. The Rio Grande failed to reach its outlet on the Gulf of Mexico for much of 2001 and 2002. The World Wildlife Fund assessed the conservation status as critical for rivers of the Southwestern Canyon region.

Mountain rivers of the Western Cordilleran region were affected by beaver trapping, mining, deforestation, tie drives, and log removal from the second half of the nineteenth century well into the twentieth century. Flow regulation and dam construction were widespread during the latter half of the twentieth century. Deforestation continues in the region, accompanied by urbanization. Water withdrawal also continues to increase, periodically attracting national attention. A recent example is the federal government's 2002 diversion of water from the Klamath River in California and Oregon to agricultural use, which resulted in the deaths of an estimated thirty-three thousand salmon and trout. The Klamath River is on the 2003 American Rivers list because of irrigation withdrawals, the presence of five hydropower dams between the agricultural basin and the coast, and pollution. The Snake River of Idaho, Washington, and Oregon is also on the list because massive hydropower dams along the river have largely destroyed a nine-hundred-mile-long spawning run historically used by an estimated two million salmon and steelhead trout. Mountain rivers of the Western Cordilleran region vary from critical conservation status in the south to endangered in the north.

Rivers of the Arctic region of the United States remain the least impacted in the country. Historical and contemporary placer mining, deforestation, and urbanization have locally altered these rivers, but the low human population density of this region has meant that the rivers have been spared some of the worst abuses. The Canning River of Alaska was on the 2002 American Rivers list, however, because proposed oil and gas exploration and development would pump millions of gallons of water from lakes in the river's delta, as well as creating extensive gravel mining along the river's floodplain. Rivers of the Arctic region have a relatively intact conservation status.

We have not cared well for our rivers. A study by Anthony Ricciardi and Joseph Rasmussen published in 1999 found that at least 123 freshwater species—79 invertebrates, 40 fish, and 4 amphibians—became extinct in North American rivers during the twentieth century. Other species of whose existence we were not even aware almost certainly went extinct as well. Where fish species are still present, they are often fewer in number, smaller in size, and younger at the time of spawning than they were historically, as chronicled for American shad by John McPhee in *The Founding Fish*. Freshwater organisms are disappearing five times faster than land animals and three times faster than coastal marine mammals, and freshwater rates of extinction are accelerating. Projected future freshwater extinction rates in North America are nearly 4 percent per decade. This rate is about the same as that for the tropical rain forests, which are considered among the most threatened ecosystems on Earth. Scientists attribute these extreme rates of extinction to both habitat degradation associated with flow regulation, river alteration, wetlands loss, and river pollution and to the widespread and frequent introduction of exotic species. More than half of the wetlands present in the United States during the 1780s had been drained by the 1980s.[1]

These extinctions impoverish a national river network that supports a greater diversity of species than most other temperate regions on the planet, thanks to North America's great heterogeneity of river habitat and the greater age of drainage networks in the geologically stable eastern United States. In *The Eternal Frontier,* Tim Flannery traces the ecological history of North America from hundreds of millions of years ago to the present. He concludes that the destruction of North America's waterways may be the greatest blow struck by European Americans at the continent's biodiversity, for the waterways represent the oldest and most distinctive biological elements on the continent. These waterways provided a critical refuge during times of crisis such as the Cretaceous-Tertiary meteorite impact sixty-five million years ago that altered climate and terrestrial ecosystems. Because of these refuges, ancient lineages such as the paddlefish or the hellbender have persisted, and North American rivers have a high diversity of insects, snails, mussels, and fish. As species after species are lost, there is the risk of whole river ecosystems unraveling.[2]

Connections

Rivers are but one manifestation of our attitude toward ourselves and the world around us. If we respect rivers, we must respect the entire world through which a river flows. That respect entails carefully considering, and taking responsibility for, each of our actions. From the food we eat, to the clothes we wear, the homes we inhabit, the work we do, and the values we espouse, our actions impact our soils, air, and water and the rivers that flow among and connect them.

I began this book hoping to increase awareness of how river ecosystems operate, and how human actions alter rivers. I thought I might be able to cite some relatively simple actions that each of us can take to contribute to preserving rivers: conserve water use, dispose of household wastes properly, work for land-use planning that preserves open space along river corridors. As I grew more involved in researching and writing this book, I realized that these actions, important as they are, will have only limited effectiveness without fundamental changes in our attitudes toward ourselves and the world of which we are a part. The society in which we live reflects immaturity, irresponsibility, and selfishness. Consider any measure you like. The food consumed by each person in the United States takes the energy equivalent of four hundred gallons of oil a year to produce, process, distribute, and prepare. Africans and Asians use about forty gallons per person for *all* their activities.[3]

How many disposable items do you use? Food containers, pens and mechanical pencils, paper, clothes, computers, appliances, cars, televisions, furniture—how many of these are composted or recycled? How many are so chemically inert that they will never leach or vaporize from landfills and contaminate soils, air, and water? Since the late 1950s, more than 750 million tons of toxic chemical wastes have been discarded in the United States. The production of these vast amounts of deadly materials is not a great mystery. These wastes come from the manufacture of our telephone poles and cell phones, our microwaves and laptops, our athletic shoes and automobiles.[4]

Between 1960 and 1990, the average American's garbage output rose 70 percent, from 2.7 pounds a day to 4.6 pounds a day, despite a tripling of recycling and composting. Much of this increase is attributable to plastic. The use of plastic increased nearly fiftyfold during these three decades, whereas the use of almost every other waste material doubled. As of 1998, average per capita garbage output dropped slightly to 4.46 pounds a day, thanks to a reduction of packaging by manufacturers and to increased composting of yard waste by communities. But we still compost or recycle only 50 percent of our paper waste, 24 percent of our plastic waste, and 12 percent of our food waste.[5]

I recently saw an advertisement proclaiming, "A bigger car: the natural product of bigger thinking." In fact, a bigger car is the product of the smallest of thinking. Such thinking treats the status symbol of the car in isolation, ignoring the enormous disproportion in resource consumption involved in manufacturing and using the car.

We do not lack in knowledge. A host of scholars and careful thinkers have described the profligacy of our ways, and better alternatives, for decades. What we lack is resolve—and the sense of urgency. As E. O.

Wilson wrote in *Consilience,* "We are drowning in information, while starving for wisdom." Anyone who cares enough to try to inform other people works from the premise that knowledge effectively disseminated produces urgency and resolve. I do not think that urgency and resolve come only from direct experience. Despite the fast pace of contemporary change, and the widespread nature of environmental degradation, the world seems newly created to each generation, and there is no perspective on cumulative loss. A child born into the Cuyahoga River valley will not know that the crooked river once teemed with fish and hosted mayflies so numerous they formed clouds of diaphanous wings when they hatched on summer evenings. She will learn of these things only from books, or from the stories told to her by earlier generations. These material and human repositories of information provide a crucial perspective that can give rise to the sense of urgency and the resolve to change our attitudes and actions. Once we reach this point, the knowledge is waiting for us.[6]

Scientists have a special responsibility in fostering a public sense of resolve and urgency. The latter twentieth century was a golden age of science in the United States. Abundant government funding supported university and agency research, and communication among scientists grew dramatically through air travel, professional journals, electronic mail, and the Internet. The resulting surge in research was closely intertwined with technological advances and with engineering. Science, engineering, and technology all make use of hypothesis proposal and testing, but I distinguish engineering and technology as being focused on manipulation, whereas science focuses on understanding interrelated processes and effects. Modern science is accused of bringing us the host of problems known as global change, including carbon dioxide–induced climate change, acid rain, desertification, and depletion of the ozone layer. Implicit in this accusation is the argument that it may be necessary to turn away from the authority of science in order to mitigate global change. I would argue that science, as opposed to engineering and technology, is our only hope for effectively addressing global change. I refer to science used as a tool for truly understanding phenomena and for recognizing interactions and outcomes that may occur beyond traditional disciplinary boundaries. In this sense, ecology is the ultimate science, for ecology integrates physics, chemistry, biology, geology, oceanography, climatology, hydrology, and other disciplines. Engineers and technicians gave us the means to cover grasslands with irrigated agricultural fields heavily dosed with fertilizers and pesticides, and to transport the resulting crops around the world. Ecologists revealed to us that these alterations have depleted and released huge reservoirs of carbon and nitrogen that contribute to carbon dioxide–induced climate change. Engineers and technicians developed atomic weapons with which to intimidate other nations. Ecologists showed that building and testing these weapons contaminate whole ecosystems with radioactive isotopes that lead to developmental abnormalities and death.

A famous physicist once remarked that science is either physics or stamp collecting. Yet the observational, exploratory science that he dismissed as stamp collecting has not only disproved fundamentally incorrect assumptions derived from physics concerning such phenomenon as the age of the Earth or the shaping of the Earth's surface by plate tectonics, but has also abundantly demonstrated the tragic shortsightedness of any scientific endeavor that does not fully consider the consequences of manipulating natural processes. Because of our narrow focus in the past, we have created an unpredictable synthetic tiger, and we are now desperately clinging to the tail. The responsibility of scientists now lies in practicing integrated science, for our society needs integrated knowledge. From this integration should flow a sense of responsibility, including a responsibility to communicate to nonscientists the extent of destruction of the natural world and the means to mitigate or reverse this destruction. In my own research, I began with a basic curiosity about how rivers function. From this work grew a respect for rivers and an urge to communicate that understanding and respect to others.

Personal Choices

How can we respect rivers? *Respect and conserve everything that goes into rivers*—everything, in other words. Eat carefully. Eat organically grown foods that are produced locally, and buy only what you will eat. Compost food scraps. Careful eating reduces the use of herbicides, insecticides, and fertilizers, as well as water use. It reduces the fuel used to transport food, and the volume of landfill taken up with discarded food and packaging. The statistics on food consumption in the United States are astonishing. A typical item of food travels fourteen hundred miles before it is eaten, changing hands at least six times en route. Just twenty years ago, that food item traveled something closer to thirty miles. It now takes ten to fifteen calories of energy to deliver one calorie of food to a U.S. consumer. And we are disproportionate in this, as in so much else. As stated earlier, each individual in the United States uses ten times more oil *on food* than the average person in Asia and Africa uses each year *for all activities.*[7]

Be aware of the sources of your food, and how it is grown. The United States uses 65 to 70 percent of its total freshwater resources to produce food and fiber for domestic needs and global exports. Irrigation is the largest consumptive water user in the United States. An estimated 137,000 million gallons per day—about 40 percent of total national freshwater use for all off-stream categories—were withdrawn for surface and groundwater supplies for irrigation during 1990. More than 85 percent of the fresh fruit and vegetables in the United States are grown with irrigation, and several are grown only with irrigation. Much of the water diverted for irrigation is not consumed and returns to downstream water sources, carrying along pesticides, nutrients, and sediment. Agriculture is the source of excess nutrients in 50 percent of the lakes and 60 percent of the river miles determined by the EPA to have impaired water quality. Choosing locally grown, unirrigated or conservatively irrigated produce changes these statistics.[8]

Avoid pesticides in your food or in your home and workplace. If you eat more than 1.2 pounds of broccoli or a quarter of a cantaloupe grown by conventional methods, you exceed your legal dose of pesticides for the year. Twenty million children younger than the age of five eat an average of eight pesticides a day. Forty suspected carcinogens appear in U.S. drinking water. Reread *Silent Spring* and consider that we use far more pesticides—one billion pounds in the United States per year—than we did in 1962. Eat foods grown without pesticides, and not only the health of your own body but the health of river ecosystems improves. And don't be too hasty in zapping the cockroach in your kitchen with poison. Household pests can be quite effectively controlled by simply plugging gaps in the floors or foundation, containing food, and keeping the house clean. Persistent bugs can be deterred with benign methods such as cedar or lavender in clothes closets, oatmeal-baited traps for silverfish, or daily hosings or soapy water sprays on infected yard plants. Nontoxic alternatives such as these are explained in *Tiny Game Hunting* by Hilary Klein and Adrian Wenner. And, to be fully effective, encourage your neighbors and communities to become pesticide-free. Although challenged by the companies TruGreen Chemlawn and Spray Tech, the Canadian Supreme Court ruled in 2001 that municipalities in Canada have the right to ban pesticide use on public and private property. As Sandra Steingraber wrote in *Living Downstream,* "a so-called private industry is engaging in a very public act when it releases toxic chemicals into a community's air, water, and soil."[9]

Use energy sparingly. Energy-efficient compact fluorescent light bulbs last six to ten times longer than incandescent bulbs. Replacing only a quarter of the most-used incandescent bulbs in your home reduces home electricity use for lighting by half, thus reducing the need for hydroelectric power and dammed rivers. And, of course, the most simple action possible: turn lights off in unused spaces. Plug air leaks from your house with caulking or weather-stripping, install storm windows, maintain heating systems by cleaning furnace filters once a month, install or upgrade insulation, and plant trees strategically around the house. Eliminate the most wasteful and polluting personal-transport items in our society: snowmobiles, personal watercraft, sport-utility vehicles, motor homes, and off-road vehicles. All of these activities reduce energy use from both hydroelectric and fossil-fuel sources, thus reducing demand for the strip-mining and mountaintop-removal that alter watersheds.[10]

Conserve water. In the United States, we each use an average of 1,400 gallons of water a day. About 100 gallons go to household use, the rest to irrigation and manufacture. Europeans use about one-third of these quantities, and people in developing countries each use an average of 12 gallons a day. Unfortunately, other people are adopting our profligate lifestyle as they adopt irrigated crops and manufactured goods. In 1900, each person in the world consumed approximately 64,000 gallons of water a year. This number climbed steadily throughout the twentieth century: 90,000 gallons in 1940, 186,000 gallons in 1970, and 210,000 by 1990. By choosing low-water faucets, toilets, and appliances and using household landscaping that does not require extensive supplemental watering, you reduce your household water consumption. Even more effective, by reducing your consumption of irrigated crops and simplifying your lifestyle, you reduce societal water consumption. As the old adage from New England expresses it: "Use it up, wear it out. Make it do, do without." More recently, the conservation ethic has been summarized as "Reduce, reuse, recycle." Clothes, housing, food, and every type of personal item, from toothpaste to snowmobiles and dishes to stereos, require energy and water to manufacture and transport. Reducing your consumption of these manufactured goods need not reduce your quality of life, but it does reduce the relentless drain on water supplies and river ecosystems.[11]

Respect life. If you respect the basic integrity and rights of all life-forms, you also inevitably respect the ecosystem processes sustaining those lives. A small incident typifies for me the attitudes of many people toward other living creatures. While working on this book, I canoed into the coastal portion of the Everglades known as the Ten Thousand Islands. I took refuge on a covered wooden platform known as a chikee during a thunderstorm. Despite the storm, the mangrove islands and sea surrounding the chikee pulsed with life. Raccoons foraged among the oyster shells at the base of the mangroves' prop roots. Fish leapt suddenly from the water as they fled larger fish. A loggerhead turtle swam briefly to the surface, inspected me, and vanished again. I was cold and wet, but the chikee seemed a privileged place in the midst of abundant life. A motorboat full of soaking-wet fishermen pulled up to the chikee in a cloud of exhaust. The men clambered stiffly up, stood about beating warmth back into their limbs, and then began fishing from the chikee. Some cast lines, others used throw nets to catch fish for bait. The shallow water around the chikee was rich with fish, and the fishermen pulled shining bodies from the dark water, throwing them into the boats to die with quivering slowness. As the storm abated, the fishermen pulled away from the chikee. They left a beer can on the platform, and small bodies floating on the oiled water.

As Edward Abbey said, *get out of your car.* Live where you work, and walk, bicycle, or use public transport as much as you can. Buy the most fuel-efficient car you can afford. Personal automobiles epitomize the wastefulness of our society, and the support network that accompanies cars—roads and gas stations—impacts tens of thousands of miles of American rivers. We spend a fortune on diet and artificial exercise in part because we have become so sedentary. The walking our grandparents and great-grandparents undertook in the normal course of daily life would initially exhaust us, their descendants. We might begin by feeling ill-used if we walked about our communities, yet we have not lost the basic ability to walk longer distances. We could choose to leave the treadmill and stationary bicycle behind and walk or ride a bicycle to run errands.

Above all else, recognize that these changes are not simply nice or conscientious choices, *they are vital decisions.* I cannot think of any description more appropriate for current health trends in the United States than that we are in the midst of a cancer epidemic. The most immediate expression is the cancers in our own bodies. How many people contracted cancer in their twenties, thirties, and forties before World War II? How many people are now finding tumors in their bodies during the prime of life? I have two good friends, active, apparently healthy women in their forties, who now live in the shadow of cancer. My parents, and the parents of my close friends, have been treated for cancer. And I know indirectly of many, many others. That we are not generally alarmed at the extent of this epidemic dismays me. I see stamps and ribbons and marathons that aim to "Race for the Cure," but I think that the nearly complete emphasis on treatment and the nearly complete denial of environmental causes of cancer is shocking. Steingraber wrote: "Our drinking water should not contain the fear of cancer. The presence of carcinogens in groundwater, no matter how faint, means we have paid too high a price for accepting the unimaginative way things are." As she died of breast cancer in her mid-fifties, Rachel Carson wrote of a fundamental but often unexpressed human right—the right to know about poisons introduced into one's environment by others, and the right to protection against them. We are paying a heavy price for our complacency as we die of cancer and watch our environment die with us.[12]

Rivers represent our future in ways that we do not yet understand. Leonardo da Vinci wrote, "In rivers, the water that you touch is the last of what has passed and the first of that which comes: so with time present." We are just realizing the implications of this as we begin to understand the global influence of river processes. Excess nitrogen from overfertilized farmlands in the Mississippi River basin flows down the river into the Gulf of Mexico, where the nutrients create a vast, oxygen-depleted dead zone that damages coastal ecosystems and fisheries. During 1999, this zone covered more than *seven thousand square miles,* an area about the size of New Jersey. The Three Gorges Dam on the Yangtze River in China may alter circulation patterns in the Sea of Japan by reducing the input of freshwater. Reducing this input by only 10 percent allows warmer, saltier, denser bottom waters to rise to the surface of the sea, warming the atmosphere over Japan and changing precipitation patterns important to Japanese agriculture. The Louisiana coastal plain is losing 0.5 to 1 percent of its land each year through subsidence and storm erosion because sediment is not being replenished by river processes. Instead, that sediment is stored behind dams, or dredged from channelized rivers and dumped elsewhere. Studies of sediment loads on the Mississippi River document an 80 percent reduction between the period before 1900 and the period from the 1960s to the 1980s. And on it goes. Present projections are that 45 percent of the world's population could be affected by either water stress or water scarcity by the year 2050, largely because of the way we alter watershed characteristics and our uses of water.[13]

We accept the idea that democracy depends on informed, involved citizens. As we grow increasingly aware of not only the political and socioeconomic but also the environmental interconnectedness of our twenty-first-century world, global survival also depends on informed, involved citizens. In describing toxic contamination of the environment, Steingraber wrote: "It is time to start pursuing alternative paths. From the right to know and the duty to inquire flows the obligation to act."[14] She wrote of going in search of our ecological roots by learning about the sources of our drinking water; the prevailing winds and what they bring us; the sources of our food; how our buildings are fumigated, our clothing cleaned, and our golf courses maintained; what our household cleaners, paints, and cosmetics contain; how our roadsides are sprayed. Rivers are an excellent place to start in the search for understanding, for they connect and integrate everything.

Collective Actions

Beyond changes in personal lifestyle, community activism and governmental and legislative changes are also necessary to protect functional, connected river ecosystems. At the community level, individuals willing to share their time and energy with environmental groups are always welcomed. Local chapters and national environmental organizations are making great strides in actions, from articulating in-stream flow requirements to increasing community awareness of the river next door. Sierra Club Water Sentinels, for example, voluntarily document the status of waterways in seven states as a means of getting the waterways cleaned up or keeping them clean. Local watershed groups lobby for more effective storm water regulations and enforcement, revitalize urban riverfronts, reevaluate existing land-use patterns, and monitor watersheds. Civic groups adopt a stream and learn about its chemistry and biology, as well as human impacts to the stream. You can research water quality in your own watershed starting with the NAWQA Web site. The Web sites for national organizations including American Rivers, the Sierra Club, and the National Audubon Society have links to many community and regional activities. If it takes a village to raise a child, it takes the citizens throughout a watershed to reconnect a river.[15]

Previous publications have outlined recommended governmental and legislative changes in detail. In their 1988 book *Down by the River,* Constance Hunt and Verne Huser call for regulation of remaining riverside habitat through the Federal Emergency Management Agency's flood insurance program and Section 404 of the Clean Water Act. They recommend federal acquisition of river corridors and the elimination of federal subsidies for agricultural irrigation, hydropower, flood insurance, and river navigation improvements by the Corps of Engineers. Restoration of physical river processes by mimicking the magnitude, timing, and variation of natural river flows could be coupled with an endangered ecosystems act to protect not only species, but entire biological communities and the landscape processes creating and maintaining the habitat on which those communities depend.[16]

The 1993 volume *Entering the Watershed* began by emphasizing that the "nation's existing riverine protection and restoration approaches and policies are inadequate and have failed." More than twenty years after the Clean Water Act and the Wild and Scenic Rivers Act, almost half of the nation's waters still fail to meet biological water-quality standards. Somewhere between one-third and three-quarters of aquatic species nationwide are rare to extinct. An estimated 70 to 90 percent of riverside vegetation has been lost or degraded. Nearly three-quarters of our rivers are impaired by flow alteration, and less than 2 percent of river miles in the United States even qualify for Wild and Scenic designation. From this grim summary, the authors of *Entering the Watershed* go on to recommend a federal initiative for community- and ecosystem-based watershed restoration.[17]

Perhaps we need a national rallying cry for rivers. Save the cottonwoods! As I write these words, national attention is focused upon international tensions and the threats of terrorism and war. Some policy-makers feel that clean water, functioning rivers, and environmental protection are luxuries not to be considered in times of national crisis. But these are not luxuries. These are literally the most fundamental components of life. Only a fool, or a foolish nation, degrades its own water. Careful, thoughtful people have recognized this for centuries. The Chinese philosopher Lao-tzu wrote: "the solution of the sage who would transform the world lies in water. Therefore when water is uncontaminated, people's hearts are upright. When water is pure, the people's hearts are at ease. The people's hearts being upright, their desires do not become dissolute and their conduct is without evil. Hence the sage, when she rules the world, does not teach the people one by one, or house by house, but takes water as the key."[18]

Our rivers are our great national heritage. They water and nourish our landscape. Even their names form our poetry. What rivers have you known, and what did their names signify? Think of them: the St. Croix, the Wailuku, the Chattahoochee, the San Joaquin. The Connecticut, Kuskowim, Potomac, Niobrara, Chickahominy, Gila, Kaskaskia, Susquehanna, Snoqualmie, Brazos, Androscoggin, and Gauley. I have known the Cuyahoga, the Rocky, the Salt, the Agua Fria, the Santa Cruz, the Colorado, the Cache la Poudre, the South Platte, and the Escalante. I hope to know many more. And I want to know them as functional, connected rivers, not as impoverished, disconnected remnants. I do not think this is too much to ask.

Notes

Chapter 3. Conquering a New World

1. B. M. Fagan, 1991, *Ancient North America: The archaeology of a continent,* Thames and Hudson, New York; D. H. Thomas, J. Miller, R. White, P. Nabokov, and P. J. DeLoria, 1993, *The Native Americans: An illustrated history,* Turner Publishing, Atlanta, Ga.

2. S. J. Pyne, 1982, *Fire in America: A cultural history of wildland and rural fire,* Princeton University Press, Princeton, N.J.

3. L. N. Shaffer, 1992, *Native Americans before 1492: The mound-building centers of the eastern woodlands,* M. E. Sharpe, Armonk, N.Y.; L. Kelly and P. G. Cross, 1984, Zooarchaeology, in C. Bareis and J. Porter, eds., *American bottom archaeology,* University of Illinois Press, Urbana; B. D. Smith, 1986, The archaeology of the southeastern United States: From Dalton to de Soto, 10,500–500 BP, *Advances in World Archaeology,* 5, 1–92; J. W. Springer, 1980, An analysis of prehistoric food remains from the Bruly St. Martin site, Louisiana, with a comparative discussion of Mississippi Valley faunal studies, *Mid-Continental Journal of Archaeology,* 5, 193–223; P. D. Royall, P. A. Delcourt, and H. R. Delcourt, 1991, Late Quaternary paleoecology and paleoenvironments of the central Mississippi alluvial valley, *Geological Society of America Bulletin,* 103, 157–70; Fagan, 1991.

4. J. Chapman, H. R. Delcourt, and P. A. Delcourt, 1989, Strawberry fields, almost forever, *Natural History,* 9/89, 50–59; Thomas et al., 1993; Pyne, 1982.

5. Thomas et al., 1993.

6. M. Sandoz, 1964, *The beaver men: Spearheads of empire,* University of Nebraska Press, Lincoln.

7. J. M. Bosch and J. D. Hewlett, 1982, A review of catchment experiments to determine the effects of vegetation changes on water yield and evapotranspiration, *Journal of Hydrology,* 55, 3–23; L. M. Reid and T. Dunne, 1984, Sediment production from forest road surfaces, *Water Resources Research,* 20, 1753–61; B. H. Heede, 1991, Increased flows after timber harvest accelerate stream disequilibrium, in *Erosion control: a global perspective,* Proceedings of the conference 22, International Erosion Control Association, Orlando, Fla., 449–54; R. F. Noss, E. T. LaRoe, and J. M. Scott, 2003, *Endangered ecosystems of the United States: A preliminary assessment of loss and degradation,* U.S. Geological Survey, Biological Resources Report, Washington, D.C.

8. M. G. Wolman, 1967, A cycle of sedimentation and erosion in urban river channels, *Geografiska Annaler,* 49A, 385–95; J. C. Knox, 1972, Valley alluviation in southwestern Wisconsin, *Annals of the Association of American Geographers,* 62, 401–10; J. C. Knox, 1977, Human impacts on Wisconsin stream channels, *Annals of the Association of American Geographers,* 67, 323–42.

9. S. W. Trimble, 1992, The Alcovy River swamps: The result of culturally accelerated sedimentation, in L. M. Dilsaver and C. E. Colten, eds., The *American envi-*

ronment: Interpretations of past geographies*, Rowman and Littlefield Publishers, Boston Way, Md.

10. R. A. Kuhnle, R. L. Bingner, and G. R. Foster, 1997, Changes in sediment load and land use on Goodwin Creek, in S. S. Y. Wang, E. J. Langendoen, and F. D. Shields, Jr., eds., *Proceedings of the Conference on Management of Landscapes Disturbed by Channel Incision, 1997*, University of Mississippi, Oxford, 375–80; G. B. Pasternack, G. S. Brush, and W. B. Hilgartner, 2001, Impact of historic land-use change on sediment delivery to a Chesapeake Bay subestuarine delta, *Earth Surface Processes and Landforms*, 26, 409–27.

11. J. R. Stilgoe, 1982, *Common landscape of America, 1580 to 1845*, Yale University Press, New Haven, Conn.

12. C. C. Watson and D. S. Biedenharn, 2000, Comparison of flood management strategies, in E. E. Wohl, ed., *Inland flood hazards: Human, riparian, and aquatic communities*, Cambridge University Press, Cambridge, U.K., 381–93.

13. A. C. Veatch, 1906, Geology and underground water resources of northern Louisiana and southern Arkansas, *U.S. Geological Survey Professional Paper 46*; J. R. Sedell and J. L. Froggatt, 1984, Importance of streamside forests to large rivers: The isolation of the Willamette River, Oregon, USA, from its floodplain by snagging and streamside forest removal, *Verhandlungen-Internationale Vereinigung für Theorelifche und Angewandte Limnologie*, 22, 1828–34; J. McPhee, 2002, *The founding fish*, Farrar, Straus and Giroux, New York; D. R. Montgomery and E. E. Wohl, 2004, Rivers and riverine landscapes, in A. R. Gillespie, S. C. Porter, and B. F. Atwater, eds. *The Quaternary period in the United States*, Elsevier, Amsterdam, 221–46; B. D. Collins, D. R. Montgomery, and A. Haas, 2002, Historic changes in the distribution and functions of large woody debris in Puget Lowland rivers, *Canadian Journal of Fisheries and Aquatic Sciences*, 59, 66–76; B. D. Collins and D. R. Montgomery, 2001, Importance of archival and process studies to characterizing pre-settlement riverine geomorphic processes and habitat in the Puget Lowland, in J. B. Dorava, D. R. Montgomery, B. Palcsak, and F. Fitzpatrick, eds., *Geomorphic processes and riverine habitat*, American Geophysical Union, Washington, D.C., 227–43; J. R. Sedell, F. H. Everest, and F. J. Swanson, 1982, Fish habitat and streamside management: Past and present, in *Proceedings of the 1981 convention of the Society of American Foresters, Sept. 27–30, 1981*, Society of American Foresters Publication 82–01, Bethesda, Md., 244–55; F. J. Triska, 1984, Role of large wood in modifying channel morphology and riparian areas of a large lowland river under pristine conditions: A historical case study, *Verhandlungen-Internationale Vereinigung für Theorelifche und Angewandte Limnologie*, 22, 1876–92.

14. Montgomery and Wohl, 2004; J. R. Sedell and W. S. Duvall, 1985, Water transportation and storage of logs, General Technical Report PNW-186, USDA Forest Service, Pacific Northwest Research Station, Portland, Ore.; Collins and Montgomery, 2001; Collins et al., 2002; E. E. Wohl, 2001, *Virtual rivers: Lessons from the mountain rivers of the Colorado Front Range*, Yale University Press, New Haven, Conn.

15. T. F. Waters, 1995, *Sediment in streams: Sources, biological effects, and control*, American Fisheries Society Monograph 7, American Fisheries Society, Bethesda, Md.; P. A. Ryan, 1991, Environmental effects of sediment on New Zealand streams: A review, *New Zealand Journal of Marine and Freshwater Research*, 25, 207–21.

16. Waters, 1995; A. D. Lemly, 1982, Modification of benthic insect communities in polluted streams: Combined effects of sedimentation and nutrient enrichment, *Hydrobiologia*, 87, 229–45; Ryan, 1991.

17. T. F. Waters, 1995, *Sediment in streams: Sources, biological effects, and control*, American Fisheries Society Monograph 7, American Fisheries Society, Bethesda, Md.

18. Ryan, 1991; Department of Natural Resources, 1970, Sheboygan River pollution investigation survey, State of Wisconsin, Madison.

19. Lemly, 1982.

20. McPhee, 2002.

21. Wohl, 2001; J. C. Frémont, 1845 (1988), *Report of the exploring expedition to the Rocky Mountains in the year 1842, and to Oregon and North California in the years 1843-'44,* Smithsonian Institution Press, Washington, D.C.

22. Wohl, 2001; M. R. Meador, 1992, Inter-basin water transfer: ecological concerns, *Fisheries* 17, 17–22.

23. Wohl, 2001.

24. F. J. Turner, 1920 (1986) *The frontier in American history,* University of Arizona Press, Tucson.

25. W. H. Goetzmann, 1986, *New lands, new men: America and the second great age of discovery,* Penguin Books, New York.

26. Goetzmann, 1986; W. H. Goetzmann, 1966, *Exploration and empire: The explorer and the scientist in the winning of the American West,* Norton and Co., New York; W. Stegner, 1954, *Beyond the hundredth meridian: John Wesley Powell and the second opening of the West,* University of Nebraska Press, Lincoln.

27. J. McPhee, 1993, *Assembling California,* Farrar, Straus and Giroux, New York.

28. M. Silva, 1986, Placer gold recovery methods, California Department of Conservation, Division of Mines and Geology, Special Publication 87.

29. Silva, 1986.

30. M. M. Hilmes and E. E. Wohl, 1995, Changes in channel morphology associated with placer mining, *Physical Geography,* 16, 223–42; L. A. James, 1991, Quartz concentration as an index of sediment mixing: Hydraulic mine-tailings in the Sierra Nevada, California, *Geomorphology,* 4, 125–44; L. A. James, 1991, Incision and morphologic evolution of an alluvial channel recovering from hydraulic mining sediment, *Geological Society of America Bulletin,* 103, 723–36; L. A. James, 1993, Sustained reworking of hydraulic mining sediment in California: G. K. Gilbert's sediment wave model reconsidered, *Zeitschrift für Geomorphologie,* 88, 49–66; L. A. James, 1989, Sustained storage and transport of hydraulic gold mining sediment in the Bear River, California, *Annals of the Association of American Geographers,* 79, 570–92; B. P. Van Haveren, 1991, Placer mining and sediment problems in interior Alaska, in S.-S. Fan and Y.-H. Kuo, eds., *Proceedings of the Fifth Federal Interagency Sedimentation Conference,* vol. 2, 10-69 to 10-74; Committee on Flood Control Alternatives in the American River Basin, 1995, *Flood risk management and the American River basin: An evaluation,* National Academy Press, Washington, D.C.; L. A. James, 1994, Channel changes wrought by gold mining: Northern Sierra Nevada, California, in *Effects of human-induced changes on hydrologic systems,* American Water Resources Association, Bethesda, Md., 629–38; K. J. Fischer and M. D. Harvey, 1991, Geomorphic response of lower Feather River to 19th century hydraulic mining operations, in *Inspiration: Come to the headwaters,* Proceedings of the 15th Annual Conference of the Association of State Floodplain Managers, Denver, Colo., 128–32; H. H. Chang, 1987, Modelling fluvial processes in streams with gravel mining, in C. R. Thorne, J. C. Bathurst, and R. D. Hey, eds., *Sediment transport in gravel-bed rivers,* Wiley and Sons, Chichester, U.K., 977–88; D. M. Bjerklie and J. D. LaPerriere, 1985, Gold-mining effects on stream hydrology and water quality, Circle Quadrangle, Alaska, *Water Resources Bulletin,* 21, 235–43; D. J. Gilvear, T. M. Waters, and A. M. Milner, 1995, Image analysis of aerial photography to quantify changes in channel morphology and instream habitat following placer mining in interior Alaska, *Freshwater Biology,* 34, 389–98.

31. H. B. C. Nitze and H. A. J. Wilkens, 1897, Gold mining in North Carolina and adjacent south Appalachian regions, *North Carolina Geological Survey Bulletin,* No. 10; McPhee, 1993.

32. C. W. Miller, Jr., 1998, *The automobile gold rushes and Depression Era mining,* University of Idaho Press, Moscow.

33. L. Ramp, 1960, Gold placer mining in southwestern Oregon, *The Ore-Bin,* 22, 75–79; E. D. Gardner and C. H. Johnson, 1934, Placer mining in the western United States, Part I, General information, hand-sluicing, and ground-sluicing, *U.S. Bureau of Mines Information Circular 6786;* T. H. Hite and G. A. Waring, 1935, Gold placer mining on Snake River in Idaho, *Economic Geology,* 30, 695–99; J. M. Kleff, 1934, Gold mining in Leadville, *Mines Magazine,* 24, 13–14; F. C. Hill, 1934, Gold mining in Oregon, *Mines Magazine,* 24, 15–16; O. A. Dingman, 1934, Placer mining in Montana, *Mines Magazine,* 24, 17–18; Miller, 1998.

34. D. J. Pisani, 1996, *Water, land, and law in the West: The limits of public policy, 1850–1920,* University Press of Kansas, Lawrence.

35. Pisani, 1996.

36. Pisani, 1996, 37.

37. C. M. Klyza, 1997, Reform at a geological pace: Mining policy on federal lands, 1964–1994, in C. Davis, ed., *Western public lands and environmental politics,* Westview Press, Boulder, Colo., 95–121.

38. Klyza, 1997.

39. Pisani, 1996; R. Solnit, 2000, The new gold rush, *Sierra,* July/August 2000, 50–57, 86; R. V. Francaviglia, 1992, Mining and landscape transformation, in L. M. Dilsaver and C. E. Colten, eds., *The American environment: Interpretations of past geographies,* Rowman and Littlefield Publishers, Boston Way, Md., 89–114; Klyza, 1997; R. Manning, 1997, *One round river: The curse of gold and the fight for the Big Blackfoot,* Henry Holt and Co., New York.

40. Waters, 1995.

41. S. B. Pentz and R. A. Kostaschuk, 1999, Effect of placer mining on suspended sediment in reaches of sensitive fish habitat, *Environmental Geology,* 37, 78–89; D. J. McLeay, I. K. Birtwell, G. F. Hartman, and G. L. Ennis, 1987, Responses of Arctic grayling (*Thymallus arcticus*) to acute and prolonged exposure to Yukon placer mining sediment, *Canadian Journal of Fisheries and Aquatic Science,* 44, 658–73; S. M. Wagener and J. D. LaPerriere, 1985, Effects of placer mining on the invertebrate communities of interior Alaska streams, *Freshwater Invertebrate Biology,* 4, 208–14; E. E. Van Nieuwenhuyse and J. D. LaPerriere, 1986, Effects of placer gold mining on primary production in subarctic streams of Alaska, *Water Resources Bulletin,* 22, 91–99; J. B. Reynolds, R. C. Simmons, and A. R. Burkholder, 1989, Effects of placer mining discharge on health and food of Arctic grayling, *Water Resources Bulletin,* 25, 625–35; J. D. LaPerriere and J. B. Reynolds, 1997, Gold placer mining and stream ecosystems of interior Alaska, in A. M. Milner and M. W. Oswood, eds., *Freshwaters of Alaska,* Springer, *Ecological Studies* 119, New York, 265–80.

42. G. K. Gilbert, 1917, Hydraulic-mining debris in the Sierra Nevada, *U.S. Geological Survey Professional Paper 105;* McPhee, 1993.

43. Meador, 1992.

44. C. N. Alpers and M. P. Hunerlach, 2000, Mercury contamination from historic gold mining in California, *U.S. Geological Survey Fact Sheet FS-061-00.*

45. Alpers and Hunerlach, 2000.

46. N. P. Prokopovich, 1984, Occurrence of mercury in dredge tailings near Fol-

som South Canal, California, *Bulletin of the Association of Engineering Geologists,* 21, 531–43.

47. R. Eisler, 1987, Mercury hazards to fish, wildlife, and invertebrates: A synoptic review, *U.S. Fish and Wildlife Service Biological Report 85 (1.10).*

48. K. J. Buhl and S. J. Hamilton, 1990, Comparative toxicity of inorganic contaminants released by placer mining to early life stages of salmonids, *Exotoxicology and Environmental Safety* 20, 325–42; J. D. LaPerriere, S. M. Wagener, and D. M. Bjerklie, 1985, Gold-mining effects on heavy metals in streams, Circle Quadrangle, Alaska, *Water Resources Bulletin,* 21, 245–52; LaPerriere and Reynolds, 1997; J. A. Stoughton and W. A. Marcus, 2000, Persistent impacts of trace metals from mining on floodplain grass communities along Soda Butte Creek, Yellowstone National Park, *Environmental Management,* 25, 305–20; D. S. Leigh, 1997, Mercury-tainted overbank sediment from past gold mining in north Georgia, USA, *Environmental Geology,* 30, 244–51.

49. D. Stiller, 2000, *Wounding the West: Montana, mining, and the environment,* University of Nebraska Press, Lincoln, 79.

50. J. N. Moore and C. Johns, 1984, Occurrence, distribution and fractionation of metals in contaminated sediments originating from mining and smelting operations along the Clark Fork River, Montana, *EOS, Transactions, American Geophysical Union,* 65, 890; Manning, 1997.

51. R. Eisler, 1988, Arsenic hazards to fish, wildlife, and invertebrates: A synoptic review, *U.S. Fish and Wildlife Service Biological Report 85 (1.12).*

52. R. Eisler, 1993, Zinc hazards to fish, wildlife, and invertebrates: A synoptic review, *U.S. Fish and Wildlife Service Biological Report 10.*

53. H. V. Leland and J. L. Carter, 1985, Effects of copper on production of periphyton, nitrogen fixation and processing of leaf litter in a Sierra Nevada, California, stream, *Freshwater Biology,* 15, 155–73; R. Eisler, 1998, Copper hazards to fish, wildlife, and invertebrates: A synoptic review, *U.S. Geological Survey USGS/BRD/BSR 1997-0002.*

54. Stiller, 2000.

55. W. A. Marcus, 1991, Managing contaminated sediments in aquatic environments: Identification, regulation, and remediation, *Environmental Law Reporter,* 21, 10020–32.

56. Marcus, 1991; I. P. G. Hutchison and R. D. Ellison, eds., 1992, *Mine waste management,* Lewis Publishers, Boca Raton, Fla., 10,025.

57. P. B. Skidmore, 1996, Placer mine reclamation at Whites Gulch: Problems and successes, *Land and Water,* September/October 1996, 14–17; Marcus, 1991.

58. W. Yeend, P. H. Stauffer, and J. W. Hendley, 1998, Rivers of gold—placer mining in Alaska, *U.S. Geological Survey Fact Sheet FS-058-98;* R. B. Wanty, B. Wang, and J. Vohden, 1997, Studies of suction dredge gold-placer mining operations along the Fortymile River, eastern Alaska, *U.S. Geological Survey Fact Sheet FS-154-97;* L. Gough, W. Day, J. Crock, B. Gamble, and M. Henning, 1997, Placer gold mining in Alaska—cooperative studies on the effect of suction dredge operations on the Fortymile River, *U.S. Geological Survey Fact Sheet FS-155-97;* T. C. Mowatt, 1987, Placer mining and surface disturbance on public lands in Alaska: Technical aspects of mitigation and reclamation, in R. F. Dworsky, ed., *Water resources related to mining and energy—preparing for the future,* Symposium Proceed., American Water Resources Association, 87-4, 453–81; R. J. Madison, 1981, Effects of placer mining on hydrologic systems in Alaska—status of knowledge, *U.S. Geological Survey Open-File Report 81-217.*

59. K. F. Karle and R. V. Densmore, 1994, Stream and floodplain restoration in a riparian ecosystem disturbed by placer mining, *Ecological Engineering*, 3, 121–33.

60. Gilbert, 1917; Wohl, 2001.

61. J. J. Hagwood, 1981, *The California Debris Commission: A history*, U.S. Army Corps of Engineers, Sacramento District, Sacramento, Calif.; A. James, 1999, Time and the persistence of alluvium: River engineering, fluvial geomorphology, and mining sediment in California, *Geomorphology*, 31, 265–90; Pyne, 1982.

62. M. J. Rohrbough, 1997, *Days of gold: The California gold rush and the American nation*, University of California Press, Berkeley; J. S. Holliday, 1999, *Rush for riches: Gold fever and the making of California*, Oakland Museum of California and University of California Press, Berkeley; E. G. Gudde, 1975, *California gold camps*, University of California Press, Berkeley.

63. W. Lindgren, 1911, Tertiary gravels of the Sierra Nevada of California, *U.S. Geological Survey Professional Paper 73;* W. E. Yeend, 1974, Gold-bearing gravel of the ancestral Yuba River, Sierra Nevada, California, *U.S. Geological Survey Professional Paper 772;* C. W. Stearn, R. L. Carroll, and T. H. Clark, 1979, *Geological evolution of North America*, 3rd ed., Wiley and Sons, New York; L. A. James, 1995, Diversion of the upper Bear River: Glacial diffluence and Quaternary erosion, Sierra Nevada, California, *Geomorphology*, 14, 131–48.

64. J. Muir, 1894 (1991), *The mountains of California*, Ten Speed Press, Berkeley, Calif.

65. Thomas et al., 1993.

66. Thomas et al., 1993; M. Sanborn, 1974, *The American: River of El Dorado*, Holt, Rinehart and Winston, New York; J. S. Holliday, 1981, *The world rushed in: The California gold rush experience*, Simon and Schuster, New York.

67. D. Goodman, 1994, *Gold seeking: Victoria and California in the 1850s*, Stanford University Press, Stanford, Calif.; C. Wollenberg, ed., 1970, *Ethnic conflict in California history*, Tinnon-Brown, Los Angeles.

68. R. Solnit, 1994, *Savage dreams: A journey into the hidden wars of the American West*, Sierra Club Books, San Francisco.

69. Sanborn, 1974; Holliday, 1981; McPhee, 1993.

70. W. H. Hall, 1880, *Report of the State Engineer to the Legislature of California, Session of 1880*, California Printing Office, Sacramento, Calif.; Solnit, 1994; Sanborn, 1974; W. H. Heuer, 1891, Improvement of San Joaquin, Mokelumne, Sacramento and Feather Rivers, Petaluma Creek, and Humboldt Harbor and Bay, California, in *Report of Major W. H. Heuer*, House Document, 52nd Congress, 1st Session, 2980–3118; G. H. Mendell, 1880, Mining debris in Sacramento River, House Document 69, 46th Congress, 2nd Session; G. H. Mendell, 1881, Protection of the navigable waters of California from injury from the debris of mines, House Document 76, 46th Congress, 3rd Session; Holliday, 1999; J. F. Mount, 1995, *California rivers and streams: The conflict between fluvial process and land use*, University of California Press, Berkeley; G. Brechin, 1999, *Imperial San Francisco: Urban power, earthly ruin*, University of California Press, Berkeley; McPhee, 1993; James, 1989; James, 1991, Quartz concentration; James, 1993; Fischer and Harvey, 1991.

71. Holliday, 1999.

72. G. K. Gilbert, 1914, Transport of debris by running water, *U.S. Geological Survey Professional Paper 86;* G. K. Gilbert, 1917, Hydraulic-mining debris in the Sierra Nevada, *U.S. Geological Survey Professional Paper 105;* James, 1989; James, 1991, Incision; James, 1993.

73. Committee on Flood Control Alternatives in the American River Basin, 1995;

L. A. James, 1997, Channel incision on the lower American River, California, from streamflow gage records, *Water Resources Research,* 33, 485–90.

74. J. Miller, R. Barr, D. Grow, P. Lechler, D. Richardson, K. Waltman, and J. Warwick, 1999, Effects of the 1997 flood on the transport and storage of sediment and mercury within the Carson River Valley, west-central Nevada, *Journal of Geology,* 107, 313–27; J. J. Rampe and D. D. Runnells, 1989, Contamination of water and sediment in a desert stream by metals from an abandoned gold mine and mill, Eureka District, Arizona, USA, *Applied Geochemistry,* 4, 445–54; Stream Solute Workshop, 1990, Concepts and methods for assessing solute dynamics in stream ecosystems, *Journal of the North American Benthological Society,* 9, 95–119; R. E. Broshears, K. E. Bencala, B. A. Kimball, and D. M. McKnight, 1993, Tracer-dilution experiments and solute-transport simulations for a mountain stream, Saint Kevin Gulch, Colorado, *U.S. Geological Survey Water-Resources Investigations Report 92-4081;* K. E. Bencala, D. M. McKnight, and G. W. Zellweger, 1990, Characterization of transport in an acidic and metal-rich mountain stream based on lithium tracer injection and simulations of transient storage, *Water Resources Research,* 26, 989–1000; B. A. Kimball, R. E. Broshears, K. E. Bencala, and D. M. McKnight, 1994, Coupling of hydrologic transport and chemical reactions in a stream affected by acid mine drainage, *Environmental Science and Technology,* 28, 2065–73; W. L. Graf, S. L. Clark, M. T. Kammerer, T. Lehman, K. Randall, and R. Schroeder, 1991, Geomorphology of heavy metals in the sediments of Queen Creek, Arizona, USA, *Catena,* 18, 567–82; S. Findlay, 1995, Importance of surface-subsurface exchange in stream ecosystems: The hyporheic zone, *Limnology and Oceanography,* 40, 159–72; K. E. Bencala, V. C. Kennedy, G. W. Zellweger, A. P. Jackman, and R. J. Avanzino, 1984, Interactions of solutes and streambed sediments, 1, An experimental analysis of cation and anion transport in a mountain stream, *Water Resources Research,* 20, 1797–1803.

75. LaPerriere et al., 1985.

76. McLeay et al., 1987; Buhl and Hamilton, 1990.

77. J. A. Stoughton and W. A. Marcus, 2000, Persistent impacts of trace metals from mining on floodplain grass communities along Soda Butte Creek, Yellowstone National Park, *Environmental Management,* 25, 305–20.

78. R. M. Yoshiyama, F. W. Fisher, and P. B. Moyle, 1998, Historical abundance and decline of chinook salmon in the Central Valley Region of California, *North American Journal of Fisheries Management,* 18, 487–521; Brechin, 1999, 50.

79. P. B. Moyle, R. M. Yoshiyama, and R. A. Knapp, 1996, Status of fish and fisheries, in *Sierra Nevada Ecosystem Project: Final report to Congress, vol. II, Assessments and scientific basis for management options,* University of California, Davis, 953–73.

80. P. B. Moyle and P. J. Randall, 1998, Evaluating the biotic integrity of watersheds in the Sierra Nevada, California, *Conservation Biology,* 12, 1318–26.

81. F. E. Price and C. E. Bock, 1983, Population ecology of the dipper (*Cinclus mexicanus*) in the Front Range of Colorado, *Studies in Avian Biology,* 7, Cooper Ornithological Society, Lawrence, Kans.; H. E. Kingery, 1996, American dipper, in *The Birds of North America,* 229; J. Dennis, 1996, *The bird in the waterfall: A natural history of oceans, rivers, and lakes,* Harper Collins Publishers, New York.

82. W. R. Goodge, 1960, Adaptations for amphibious vision in the dipper (*Cinclus mexicanus*), *Journal of Morphology,* 107, 79–91; Kingery, 1996.

83. J. O. Sullivan, 1973, *Ecology and behavior of the dipper, adaptations of a passerine to an aquatic environment,* unpublished PhD dissertation, University of Montana, Missoula; Price and Bock, 1983.

84. G. J. Bakus, 1957, *The life history of the dipper on Rattlesnake Creek, Missoula*

County, Montana, unpublished MA thesis, Montana State University, Bozeman; Price and Bock, 1983.

85. Bakus, 1957.

86. Kingery, 1996.

87. J. Muir, 1894 (1991), *The mountains of California*, Ten Speed Press, Berkeley, Calif., 283.

88. Bakus, 1957.

89. S. M. Strom, 2000, *The utility of metal biomarkers in assessing the toxicity of metals in the American dipper (Cinclus mexicanus)*, unpublished MS thesis, Colorado State University, Ft. Collins.

90. Strom, 2000, 12.

91. Moyle and Randall, 1998.

Chapter 4. Poisoning America

1. F. J. Turner, 1920 (1986), *The frontier in American history*, University of Arizona Press, Tucson; K. Warner and I. R. Porter, 1960, Experimental improvement of a bulldozed trout stream in northern Maine, *Transactions of the American Fisheries Society*, 89, 59–63.

2. M. G. Wolman and A. P. Schick, 1967, Effects of construction on fluvial sediment, urban and suburban areas of Maryland, *Water Resources Research*, 3, 451–64; D. G. Anderson, 1970, Effects of urban development on floods in northern Virginia, *U.S. Geological Survey Water-Supply Paper 2001-C;* A. Chin and K. J. Gregory, 2001, Urbanization and adjustment of ephemeral stream channels, *Annals of the Association of American Geographers*, 91, 595–608; B. P. Bledsoe and C. C. Watson, 2001, Effects of urbanization on channel instability, *Journal of the American Water Resources Association*, 37, 255–70; S. W. Trimble, 1997, Contribution of stream channel erosion to sediment yield from an urbanizing watershed, *Science*, 278, 1442–44.

3. A. Vileisis, 1999, Cash register rivers: Waterways lost to private profit, in J. Scherff, ed., *The piracy of America: Profiteering in the public domain*, Clarity Press, Atlanta, Ga., 49–66; J. R. Stilgoe, 1982, *Common landscape of America, 1580 to 1845*, Yale University Press, New Haven, Conn.; H. D. Thoreau, 1849 (1983), *A week on the Concord and Merrimack rivers*, Princeton University Press, Princeton, N.J.; J. McPhee, 2002, *The founding fish*, Farrar, Straus and Giroux, New York.

4. G. Hardin, 1968, The tragedy of the commons, *Science*, 162, 1243–48.

5. D. W. Meinig, 1998, *Transcontinental America, 1850–1915. The shaping of America: A geographical perspective on 500 years of history*, vol. 3, Yale University Press, New Haven, Conn.

6. M. B. Bogue, 2000, *Fishing the Great Lakes: An environmental history, 1783–1933*, University of Wisconsin Press, Madison.

7. R. W. Sellars, 1997, *Preserving nature in the national parks: A history*, Yale University Press, New Haven, Conn.

8. W. M. Davis, 1899, The peneplain, *American Geologist*, 23, 207–39; W. M. Davis, 1905, Complications of the geographical cycle, *Report of the 8th Geographical Congress 1904*, 150–63.

9. B. L. Johnson, W. B. Richardson, and T. J. Naimo, 1995, Past, present, and future concepts in large river ecology, *BioScience*, 45, 134–41; G. W. Minshall, K. W. Cummins, R. C. Petersen, C. E. Cushing, D. A. Bruns, J. R. Sedell, and R. L. Vannote, 1985, Developments in stream ecosystem theory, *Canadian Journal of Fisheries and Aquatic Sciences*, 42, 1045–55; D. E. Brown, W. L. Minckley, and J. P. Collins, 1994,

Historical background to southwest ecological studies, in D. E. Brown, ed., *Biotic communities: Southwestern United States and northwestern Mexico,* University of Utah Press, Salt Lake City, 17–23; A. N. Strahler, 1992, Quantitative/dynamic geomorphology at Columbia 1945–1960: A retrospective, *Progress in Physical Geography,* 16, 65–84; J. D. Vitek and D. F. Ritter, 1993, Geomorphology in the USA, in H. J. Walker and W. E. Grabau, eds., *The evolution of geomorphology,* Wiley and Sons, Chichester, U.K., 469–81.

10. Minshall et al., 1985; N. Hynes, 1970, *The ecology of running waters,* University of Toronto Press, Toronto, Canada; H. B. N. Hynes, 1975, The stream and its valley, *Verhandlungen-Internationale Vereinigung für Theorelifche und Angewandte Limnologie,* 19, 1–15; R. L. Vannote, G. W. Minshall, K. W. Cummins, J. R. Sedell, and C. E. Cushing, 1980, The river continuum concept, *Canadian Journal of Fisheries and Aquatic Sciences,* 37, 130–37.

11. W. J. Junk, P. B. Bayley, and R. E. Sparks, 1989, The flood pulse concept in river-floodplain systems, *Can. Spec. Publ. Fisheries and Aquatic Sciences,* 106, 110–27.

12. Minshall et al., 1985; J. W. Elwood, J. D. Newbold, R. V. O'Neill, and W. Van Winkle, 1983, Resource spiraling: An operational paradigm for analyzing lotic ecosystems, in T. D. Fontaine and S. M. Bartell, eds., *Dynamics of lotic ecosystems,* Ann Arbor Science, Ann Arbor, Mich., 3–27; J. R. Webster and B. C. Patten, 1979, Effects of watershed perturbation on stream potassium and calcium dynamics, *Ecological Monographs,* 49, 51–72.

13. T. F. Waters, 1995, *Sediment in streams: Sources, biological effects, and control,* American Fisheries Society Monograph 7, American Fisheries Society, Bethesda, Md.

14. C. E. Colten, 1992, Illinois River pollution control, 1900–1970, in L. M. Dilsaver and C. E. Colten, eds., *The American environment: Interpretations of past geographies,* Rowman and Littlefield Publishers, Boston Way, Md.; S. A. Forbes, 1910, The investigation of a river system in the interest of its fisheries, *Transactions of the American Fisheries Society,* 40, 179–93, 179.

15. Colten, 1992; R. E. Sparks, 1984, The role of contaminants in the decline of the Illinois River: Implications for the upper Mississippi, in J. G. Wiener, R. V. Anderson, and D. R. McConville, eds., *Contaminants in the Upper Mississippi River,* Proceed. of the 15th Annual Meeting of the Mississippi River Research Consortium, Butterworth Publishers, Stoneham, Mass.

16. Colten, 1992.

17. N. G. Bhowmik and M. Demissie, 2002, *Summary of research on the Illinois River and Peoria Lake by the Illinois State Water Survey related to sedimentation and water level fluctuations,* Illinois State Water Survey, Champaign; N. G. Bhowmik and M. Demissie, 2001, *River geometry, bank erosion, and sand bars within the main stem of the Kankakee River in Illinois and Indiana,* Illinois State Water Survey, Contract Report 2001-09; N. G. Bhowmik et al., 1994, The 1993 flood on the Mississippi River in Illinois, *Illinois State Water Survey Miscellaneous Publication 151.*

18. N. G. Bhowmik and M. Demissie, 2002, *Potential island construction sites within the lower Peoria Lake,* Illinois State Water Survey, Champaign.

19. Ohio River Valley, 1962, *Aquatic-life resources on the Ohio River,* Ohio River Valley Water Sanitation Commission, Cincinnati.

20. Ohio River Valley, 1962.

21. Department of Natural Resources, 1970, *Sheboygan River pollution investigation survey,* State of Wisconsin, Madison; Department of the Army, 1971, *Cuyahoga River basin, Ohio restoration study,* First Interim Report, Buffalo District, Corps of Engineers, Buffalo, N.Y.; Department of Natural Resources, 1971, *Duck Creek-Pensaukee*

River pollution investigation survey, Department of Natural Resources, State of Wisconsin, Madison; J. R. E. Jones, 1964, *Fish and river pollution,* Butterworths, London; G. A. Best and S. L. Ross, 1977, *River pollution studies,* Liverpool University Press, Liverpool, U.K.

22. Department of the Army, 1971.

23. Department of the Army, 1971.

24. S. R. Rod, 1989, *Estimation of historical pollution trends using mass balance principles: Selected metals and pesticides in the Hudson-Raritan basin, 1880 to 1980,* unpublished PhD dissertation, Carnegie Mellon University, Pittsburgh, Pa.

25. Rod, 1989.

26. A. W. Sweeton, 1978, *PCB and the Housatonic River: A review and recommendations,* Connecticut Academy of Science and Engineering, Hartford.

27. Sweeton, 1978; R. Eisler, 1986, Polychlorinated biphenyl hazards to fish, wildlife, and invertebrates: A synoptic review, *U.S. Fish and Wildlife Service Biological Report 85 (1.7).*

28. Sweeton, 1978.

29. Eisler, 1986.

30. P. Rauber, 2001, Fishing for life, *Sierra,* 86, 42–49; Sweeton, 1978.

31. S. Steingraber, 1997, *Living downstream: An ecologist looks at cancer and the environment,* Addison-Wesley Publishing Co., Reading, Mass.

32. Public Health Service, 1965, United States of America, in J. Litwin, ed., *Control of river pollution by industry,* International Institute of Administrative Sciences, Brussels, Belgium, 159–67.

33. Vileisis, 1999.

34. Vileisis, 1999; T. Palmer, 1994, *Lifelines: The case for river restoration,* Island Press, Washington, D.C.; B. Doppelt, M. Scurlock, C. Frissell, and J. Karr, 1993, *Entering the watershed: A new approach to save America's river ecosystems,* Island Press, Washington, D.C.

35. K. C. Rice, 1999, Trace-element concentrations in streambed sediments across the conterminous United States, *Environmental Science and Technology,* 33, 2499–04; G. R. Wall, K. Riva-Murray, and P. J. Phillips, 1998, Water quality in the Hudson River basin, New York and adjacent states, 1992–1995, *U.S. Geological Survey Circular 1165;* R. Eisler, 1998, Copper hazards to fish, wildlife, and invertebrates: A synoptic review, *U.S. Geological Survey USGS/BRD/BSR 1997-0002.*

36. Rice, 1999, 2499.

37. R. Eisler, 1985, Cadmium hazards to fish, wildlife, and invertebrates: A synoptic review, *U.S. Fish and Wildlife Service Biological Report 85 (1.2).*

38. Eisler, 1985.

39. R. Eisler, 1986. Chromium hazards to fish, wildlife, and invertebrates: A synoptic review, *U.S. Fish and Wildlife Service Biological Report 85 (1.6).*

40. R. Eisler, 1988, Lead hazards to fish, wildlife, and invertebrates: A synoptic review, *U.S. Fish and Wildlife Service Biological Report 85 (1.14);* S. M. Strom, 2000, *The utility of metal biomarkers in assessing the toxicity of metals in the American dipper (Cinclus mexicanus),* unpublished MS thesis, Colorado State University, Ft. Collins.

41. R. Eisler, 1993, Zinc hazards to fish, wildlife, and invertebrates: A synoptic review, *U.S. Fish and Wildlife Service Biological Report 10.*

42. L. H. Nowell, P. D. Capel, and P. D. Dileanis, 1999, Pesticides in stream sediment and aquatic biota—distribution, trends, and governing factors, *Pesticides in the hydrologic system series,* vol. 4. CRC Press, Boca Raton, Fla.

43. L. Nowell, 2001, *Organochlorine pesticides and PCBs in bed sediment and aquatic*

biota from United States rivers and streams: Summary statistics, preliminary results of the National Water Quality Assessment Program (NAWQA), 1992–1998, NAWQA Pesticide National Synthesis Project, Web site http://ca.water.usgs.gov/pnsp/bst2; Natural Resources Defense Council Web site, 2003, *http://www.nrdc.org/breastmilk/chem2.asp;* Nowell et al., 1999.

44. U.S. Geological Survey, 1999, The quality of our nation's waters: Nutrients and pesticides, *U.S. Geological Survey Circular 1225.*

45. U.S. Geological Survey, 1999.

46. U.S. Geological Survey, 1999, 19, 6, 6–7.

47. U.S. Geological Survey, 1999.

48. R. Eisler, 1989, Atrazine hazards to fish, wildlife, and invertebrates: A synoptic review, *U.S. Fish and Wildlife Service Biological Report 85 (1.18);* R. Eisler, 1990, Chlordane hazards to fish, wildlife, and invertebrates: A synoptic review, U.*S. Fish and Wildlife Service Biological Report 85 (1.21).*

49. Eisler, 1989.

50. Eisler, 1990.

51. T. Colborn, D. Dumanoski, and J. P. Myers, 1997, Hormone impostors, *Sierra,* 82, 28–35; T. Colborn and R. A. Liroff, 1990, Toxics in the Great Lakes, *EPA Journal,* 16, 5–8.

52. Colborn et al., 1997; E. Carlsen, A. Giwercman, N. Keiding, and N. E. Skakkebaek, 1995, Declining semen quality and increasing incidence of testicular cancer: Is there a common cause?, *Environmental Health Perspectives,* 103, 137–39; A. Pathomvanich, D. P. Merke, and G. P. Chrousos, 2000, Early puberty: A cautionary tale, *Pediatrics,* 105, 115–16; T. Colborn and K. Thayer, 2000, Aquatic ecosystems: Harbingers of endocrine disruption, *Ecological Applications,* 10, 949–57; T. Colborn, M. Smolen, and R. Rolland, 1996, Taking a lead from wildlife, *Neurotoxicology and Teratology,* 18, 235–37; S. L. Goodbred, R. J. Gilliom, T. S. Gross, N. P. Denslow, W. L. Bryant, and T. R. Schoeb, 1997, Reconnaissance of 17β-estradiol, 11-ketotestosterone, vitellogenin, and gonad histopathology in common carp of United States streams: Potential for contaminant-induced endocrine disruption, *U.S. Geological Survey Open-File Report 96-627;* T. Colborn, F. S. Vom Saal, and A. M. Soto, 1994, Developmental effects of endocrine-disrupting chemicals in wildlife and humans, *Environmental Impact Assessment Review,* 14, 469–89; S. B. Smith, R. J. Sloan, and B. P. Baldigo, 2000, Altered endocrine biomarkers in selected fish species in the Hudson River, New York, *U.S. Geological Survey Fact Sheet FS-113-00.*

53. T. Colborn, 1994, The wildlife/human connection: Modernizing risk decisions, *Environmental Health Perspectives,* 102, 55–59; B. C. Gladen, T. D. Zadorozhnaja, N. Chislovska, D. O. Hrykorczuk, M. C. Kennicutt, and R. E. Little, 2000, Polycyclic aromatic hydrocarbons in placenta, *Human and Experimental Toxicology,* 19, 597–603; H. M. Blanck, M. Marcus, V. Hertzberg, P. E. Tolbert, C. Rubin, A. K. Henderson, and R. H. Zhang, 2000, Determinants of polybrominated biphenyl serum decay among women in the Michigan PBB cohort, *Environmental Health Perspectives,* 108, 147–52; P. Short and T. Colborn, 1999, Pesticide use in the U.S. and policy implications: A focus on herbicides, *Toxicology and Industrial Health,* 15, 240–75.

54. Colborn et al., 1997.

55. T. Colborn, 1995, Pesticides—how research has succeeded and failed to translate science into policy: Endocrinological effects on wildlife, *Environmental Health Perspectives,* 103, 81–86.

56. U.S. Geological Survey, 1999.

57. Steingraber, 1997.

58. U.S. Geological Survey, 1999; E. A. Graffy, D. R. Helsel, and D. K. Mueller, 1996, Nutrients in the nation's waters: Identifying problems and progress, *U.S. Geological Survey Fact Sheet FS-218-96.*

59. D. E. Fisher and M. J. Fisher, 2001, The nitrogen bomb, *Discover,* 22, 51–57.

60. B. T. Nolan and J. D. Stoner, 2000, Nutrients in groundwaters of the conterminous United States, 1992–1995, *Environmental Science and Technology,* 34, 1156–65.

61. Nolan and Stoner, 2000; D. K. Mueller and D. R. Helsel, 1996, Nutrients in the nation's waters—too much of a good thing?, *U.S. Geological Survey Circular 1136;* G. M. Clark, D. K. Mueller, and M. A. Mast, 2000, Nutrient concentrations and yields in undeveloped stream basins of the United States, *Journal of the American Water Resources Association,* 36, 849–60; B. J. Peterson, W. M. Wollheim, P. J. Mulholland, J. R. Webster, J. L. Meyer, J. L. Tank, E. Marti, W. B. Bowden, H. M. Valett, A. E. Hershey, W. H. McDowell, W. K. Dodds, S. K. Hamilton, S. Gregory, and D. D. Morall, 2001, Control of nitrogen export from watersheds by headwater streams, *Science,* 292, 86–90.

62. U.S. Geological Survey, 1999; Nolan and Stoner, 2000.

63. National Science and Technology Council, 2000, *Integrated assessment of hypoxia in the northern Gulf of Mexico,* Committee on Environment and Natural Resources, National Science and Technology Council, Washington, D.C.; D. A. Goolsby, 2000, Mississippi Basin nitrogen flux believed to cause Gulf hypoxia, *EOS, Transactions, American Geophysical Union,* 81, 321–27; W. J. Mitsch, J. W. Day, Jr., J. W. Gilliam, P. M. Groffman, D. L. Hey, G. W. Randall, and N. Wang, 2001, Reducing nitrogen loading to the Gulf of Mexico from the Mississippi River basin: Strategies to counter a persistent ecological problem, *BioScience,* 51, 373–88; M. E. Borsuk, C. A. Stow, and K. H. Reckhow, 2003, Integrated approach to total maximum daily load development for Neuse River estuary using Bayesian probability network model, *Journal of Water Resources Planning and Management,* 129, 271–82; C. A. Stow, M. E. Borsuk, and D. W. Stanley, 2001, Long-term changes in watershed nutrient inputs and riverine exports in the Neuse River, North Carolina, *Water Research,* 35, 1489–99.

64. H. Brass et al., 1977, The national organics monitoring survey: Samplings and analyses for purgeable organic compounds, in R. B. Pojasek, ed., *Drinking water quality enhancement through source protection,* Ann Arbor Science Publishers, Ann Arbor, Mich., 238–49; J. M. Symons et al., 1975, National organics reconnaissance survey for halogenated organics, *Journal of AWWA,* 67, 634–47; Steingraber, 1997, 204.

65. Fisher and Fisher, 2001.

66. Mueller and Helsel, 1996, 7.

67. R. E. Rathbun, 1998, Transport, behavior, and fate of volatile organic compounds in streams, *U.S. Geological Survey Professional Paper 1589;* Steingraber, 1997.

68. P. J. Squillace, M. J. Moran, W. W. Lapham, C. V. Price, R. M. Clawges, and J. S. Zogorski, 1999, Volatile organic compounds in untreated ambient groundwater of the United States, 1985–1995, *Environmental Science & Technology,* 33, 4176–87; T. J. Lopes and D. A. Bender, 1998, Nonpoint sources of volatile organic compounds in urban areas—relative importance of urban land surfaces and air, *Environmental Pollution,* 101, 221–30.

69. P. J. Squillace, 1999, *MTBE in the nation's ground water,* U.S. Geological Survey NAWQA Web site: *http://sd.water.usgs.gov/nawqa/vocns/mtbe.html* Rathbun, 1998.

70. Graffy et al., 1996.

71. Steingraber, 1997; Centers for Disease Control and Prevention, 2001, *Na-*

tional report on human exposure to environmental chemicals, Centers for Disease Control and Prevention, National Center for Environmental Health, NCEH Publication No. 01-0164.; V. McGovern, 2004, Sport-caught fish and breast cancer: Angling for more data, *Environmental Health Perspectives,* 112, A112.

72. Steingraber, 1997, 73.

73. Steingraber, 1997, 139.

74. F. A. Fitzpatrick, T. L. Arnold, and J. A. Colman, 1998, Surface-water-quality assessment of the Upper Illinois River basin in Illinois, Indiana, and Wisconsin—spatial distribution of geochemicals in the fine fraction of streambed sediment, 1987, *U.S. Geological Survey Water-Resources Investigations Report 98-4109;* J. V. Klump, J. T. Waples, D. N. Edgington, and K. A. Orlandini, 2002, Tracking sediment resuspension and transport in the nearshore of the Great Lakes, abstract in *U.S.–Chinese Joint Workshop on Sediment Transportation and Environmental Studies,* Marquette University, Milwaukee, Wisc., July 2002, 33.

75. K. A. Gray, J. A. Kostel, and A. St. Amand, 2002, The impact of metal and organic contaminants on the structure of periphyton in lotic sediments: Observations at various scales, abstract in *U.S.–Chinese Joint Workshop on Sediment Transportation and Environmental Studies,* Marquette University, Milwaukee, Wisc., July 2002, 25.

76. Personal communication, T. Wentland, Wisconsin Department of Natural Resources, July 2002.

77. Personal communication, T. Wentland, Wisconsin Department of Natural Resources, July 2002.

78. Personal communication, T. Wentland, Wisconsin Department of Natural Resources, July 2002.

79. G. D. Casey, D. N. Myers, D. P. Finnegan, and M. E. Wieczorek, 1998, National water-quality assessment of the Lake Erie-Lake St. Clair Basin, Michigan, Indiana, Ohio, Pennsylvania and New York—environmental and hydrologic setting, *U.S. Geological Survey Water-Resources Investigations Report 97-4256.*

80. Casey et al., 1998; R. Eisler, 1987, Polycyclic aromatic hydrocarbon hazards to fish, wildlife, and invertebrates: A synoptic review, *U.S. Fish and Wildlife Service Biological Report 85 (1.11).*

81. Casey et al., 1998.

82. Casey et al., 1998.

83. E. Pianin, 2001, Bush wants to give environmental oversight to states, *The Coloradoan,* July 23, 2001, A7.

84. Casey et al., 1998.

85. T. A. Edsall and C. N. Raphael, 1988, The St. Clair River and Lake St. Clair, Michigan: An ecological profile, *U.S. Fish and Wildlife Service Biological Report 85 (7.3).*

86. Edsall and Raphael, 1988.

87. The following discussion on hellbenders is based on J. H. Harding, 1997, *Amphibians and reptiles of the Great Lakes region,* University of Michigan Press, Ann Arbor.

88. R. W. Guimond and V. H. Hutchison, 1973, Aquatic respiration: An unusual strategy in the hellbender *Cryptobranchus alleganiensis* (Daudin), *Science,* 182, 1263–65.

89. E. Routman, R. Wu, and A. R. Templeton, 1994, Parsimony, molecular evolution, and biogeography: The case of the North American giant salamander, *Evolution,* 48, 1799–1809.

90. J. Danch, 1996, The hellbender, *Reptiles,* 4, 48–59.

Chapter 5. Institutional Conquest

1. S. A. Changnon, 1998, The historical struggle with floods on the Mississippi River basin: Impacts of recent floods and lessons for future flood management and policy, *Water International,* 23, 263–71; Committee on Government Operations, 1973, *Stream channelization: What federally financed draglines and bulldozers do to our nation's streams,* Fifth report by the Committee on Government Operations together with additional views, U.S. Government Printing Office, Washington, D.C.

2. W. Stegner, 1953, *Beyond the 100th meridian: John Wesley Powell and the second opening of the West,* University of Nebraska Press, Lincoln.

3. J. M. Barry, 1997, *Rising Tide: The great Mississippi flood of 1927 and how it changed America,* Simon & Schuster, New York.

4. P. W. Simpson, J. R. Newman, M. A. Keirn, R. M. Matter, and P. A. Guthrie, 1982, *Manual of stream channelization impacts on fish and wildlife,* FWS/OBS-82/24.

5. Simpson et al., 1982.

6. Committee on Government Operations, 1973; C. C. Watson and D. S. Biedenharn, 2000, Comparison of flood management strategies, in E. E. Wohl, ed., *Inland flood hazards: Human, riparian and aquatic communities,* Cambridge University Press, Cambridge, U.K., 381–93; A. Vileisis, 1999, Cash register rivers: Waterways lost to private profit, in J. Scherff, ed., *The piracy of America: Profiteering in the public domain,* Clarity Press, Atlanta, Ga., 49–66.

7. W. L. Graf, 1999, Dam nation: A geographic census of American dams and their large-scale hydrologic impacts, *Water Resources Research,* 35, 1305–11.

8. Graf, 1999.

9. Committee on Government Operations, 1973, 3–4; C. E. Hunt and V. Huser, 1988, *Down by the river: The impact of federal water projects and policies on biological diversity,* Island Press, Washington, D.C.

10. S. E. Bunn and A. H. Arthington, 2002, Basic principles and ecological consequences of altered flow regimes for aquatic biodiversity, *Environmental Management,* 30, 492–507; C. Nilsson and M. Svedmark, 2002, Basic principles and ecological consequences of changing water regimes: Riparian plant communities, *Environmental Management,* 30, 468–80; N. L. Poff and D. D. Hart, 2002, How dams vary and why it matters for the emerging science of dam removal, *BioScience,* 52, 659–68; K. D. Fausch, C. T. Torgersen, C. V. Baxter, and H. W. Li, 2002, Landscapes to riverscapes: Bridging the gap between research and conservation of stream fishes, *BioScience,* 52, 483–98; E. Goldsmith and N. Hildyard, 1984, *The social and environmental effects of large dams.* Sierra Club Books, San Francisco; G. P. Williams and M. G. Wolman, 1984, Effects of dams and reservoirs on surface-water hydrology: Changes in rivers downstream from dams, *U.S. Geological Survey Professional Paper 1286.*

11. W. E. Stegner, 1974, *The uneasy chair: A biography of Bernard de Voto,* Doubleday, Garden City, N.Y.

12. Stegner, 1974; Graf, 1999.

13. T. Dunne and L. B. Leopold, 1978, *Water in environmental planning,* W. H. Freeman, San Francisco; Glen Canyon Institute Web site, 2003, http://www.glencanyon.org.

14. M. Collier, R. H. Webb, and J. C. Schmidt, 1996, Dams and rivers: Primer on the downstream effects of dams, *U.S. Geological Survey Circular 1126.*

15. Collier et al., 1996; R. H. Webb, P. T. Pringle, S. L. Reneau, and G. R. Rink, 1988, Monument Creek debris flow, 1984: Implications for formation of rapids on

the Colorado River in Grand Canyon National Park, *Geology,* 16, 50–54; S. W. Kieffer, 1989, Geologic nozzles, *Reviews of Geophysics,* 27, 3–38; S. M. Wiele, J. B. Graf, and J. D. Smith, 1996, Sand deposition in the Colorado River in the Grand Canyon from flooding of the Little Colorado River, *Water Resources Research,* 32, 3579–96.

16. A. D. Konieczki, J. B. Graf, and M. C. Carpenter, 1997, Streamflow and sediment data collected to determine the effects of a controlled flood in March and April 1996 on the Colorado River between Lees Ferry and Diamond Creek, Arizona, *U.S. Geological Survey Open-File Report 97-224.*

17. G. E. Galloway, 1994, *Sharing the challenge: Floodplain management into the 21st century,* Report of the Interagency Floodplain Management Review Committee, Washington, D.C.; C. B. Belt, Jr., 1975, The 1973 flood and man's constriction of the Mississippi River, *Science,* 189, 681–84; D. Jehl, 2001, Mississippi floods revive debate on what government should do, *New York Times,* April 27, 2001, A1, A16; Changnon, 1998.

18. R. Gillette, 1972, Stream channelization: Conflict between ditchers, conservationists, *Science,* 176, 890–94; R. Schoof, 1980, Environmental impact of channel modification, *Water Resources Bulletin,* 16, 697–701.

19. Committee on Government Operations, 1973, 3; D. L. Henegar and K. W. Harmon, 1971, A review of references to channelization and its environmental impact, in E. Schneberger and J. L. Funk, eds., *Stream channelization: A symposium,* American Fisheries Society Special Publication No. 2, Omaha, Nebr., 20–28.

20. R. C. Heidinger, 1989, Fishes in the Illinois portion of the upper Des Plaines River, *Trans. Illinois Academy of Sciences,* 82, 85–96; C. J. Barstow, 1971, Impact of channelization on wetland habitat in the Obion-Forked Deer Basin, Tennessee, in E. Schneberger and J. L. Funk, eds., *Stream channelization: A symposium,* American Fisheries Society Special Publication No. 2, Omaha, Nebr., 20–28.

21. D. Shankman, 1993, Channel migration and vegetation patterns in the southeastern coastal plain, *Conservation Biology,* 7, 176–83.

22. S. A. Schumm, M. D. Harvey, and C. C. Watson, 1984, *Incised channels: Morphology, dynamics and control,* Water Resources Publications, Littleton, Colo.; A. Simon, 1994, Gradation processes and channel evolution in modified west Tennessee streams: Process, response, and form, *U.S. Geological Survey Professional Paper 1470;* C. R. Hupp, 1992, Riparian vegetation recovery patterns following stream channelization: A geomorphic perspective, *Ecology,* 7, 1209–26; B. L. Rhoads, 1990, The impact of stream channelization on the geomorphic stability of an arid-region river, *National Geographic Research,* 6, 157–77; A. Simon and S. Darby, 1999, The nature and significance of incised river channels, in S. E. Darby and A. Simon, eds., *Incised river channels: Processes, forms, engineering and management,* Wiley and Sons, Chichester, U.K., 3–18; C. R. Hupp, 1999, Relations among riparian vegetation, channel incision processes and forms, and large woody debris, in S. E. Darby and A. Simon, eds., *Incised river channels: Processes, forms, engineering and management,* Wiley and Sons, Chichester, U.K., 219–45; R. S. Barnard and W. N. Melhorn, 1982, Morphologic and morphometric response to channelization: The case history of Big Pine Creek Ditch, Benton County, Indiana, in R. G. Craig and J. L. Craft, eds., *Applied geomorphology,* Allen and Unwin, London, 224–39; P. J. Terrio and J. E. Nazimek, 1997, Changes in cross-sectional geometry and channel volume in two reaches of the Kankakee River in Illinois, 1959–1994, *U.S. Geological Survey Water-Resources Investigations Report 96-4261.*

23. A. Simon and C. H. Robbins, 1987, Man-induced gradient adjustment of the South Fork Forked Deer River, west Tennessee, *Environmental Geology and Water Sci-*

ence, 9, 109–18; R. H. Kesel and E. G. Yodis, 1992, Some effects of human modifications on sand-bed channels in southwestern Mississippi, U.S.A., *Environmental Geology and Water Science*, 20, 93–104; D. M. Patrick, L. M. Smith, and C. B. Whitten, 1982, Methods for studying accelerated fluvial change, in R. D. Hey et al., eds., *Gravel-bed rivers*, Wiley and Sons, Chichester, U.K., 783–815.

24. E. A. Keller, 1977, Pools, riffles, and channelization, *Environmental Geology*, 2, 119–27.

25. E. H. Grissinger and J. B. Murphey, 1982, Present "problem" of stream channel instability in the Bluff Area of northern Mississippi, *Journal of the Mississippi Academy of Sciences*, 28, 117–28.

26. M. W. Doyle and F. D. Shields, Jr., 1998, Perturbations of stage hydrographs caused by channelization and incision, *International Water Resources Engineering Conference Proceedings*, American Society of Civil Engineers, 1, 736–41; N. Pinter, R. Thomas, and J. H. Wlosinski, 2000, Regional impacts of levee construction and channelization, middle Mississippi River, USA, in J. Marsalek, W. E. Watt, E. Zeman, and F. Sieker, eds., *Flood issues in contemporary water management*, Kluwer Academic Publishers, Dordrecht, Netherlands, 351–61; D. Shankman and S. A. Samson, 1991, Channelization effects on Obion River flooding, western Tennessee, *Water Resources Bulletin*, 27, 247–54; C. H. Robbins and A. Simon, 1983, Man-induced channel adjustment in Tennessee streams, *U.S. Geological Survey Water-Resources Investigations Report 82*, 4098; D. Shankman and T. B. Pugh, 1992, Discharge response to channelization of a coastal plain stream, *Wetlands*, 12, 157–62.

27. S. S. Hahn, 1982, Stream channelization: Effects on stream fauna, in P. E. Greeson, ed., Biota and biological principles of the aquatic environment, *U.S. Geological Survey Circular 848-A*, A43–A49; C. E. Petersen, 1991, Water quality of the West Branch of the Dupage River and Kline Creek, Illinois, as evaluated using the arthropod fauna and chemical measurements, *The Great Lakes Entomologist*, 24, 127–31.

28. Hahn, 1982; T. H. Diehl, 1997, Drift in channelized streams, in S. S. Y. Wang, E. J. Langendoen, and F. D. Shields, Jr., eds., *Proceedings of the Conference on Management of Landscapes Disturbed by Channel Incision, 1997*, University of Mississippi, Oxford, 35–41.

29. D. L. Scarnecchia, 1988, The importance of streamlining in influencing fish community structure in channelized and unchannelized reaches of a prairie stream, *Regulated Rivers: Research and Management*, 2, 155–66.

30. Hahn, 1982; B. L. Griswold, C. Edwards, L. Woods, and E. Weber, 1978, *Some effects of stream channelization on fish populations, macroinvertebrates, and fishing in Ohio and Indiana*, FWS/OBS-77/46; C. B. Portt, E. K. Balon, and D. L. G. Noakes, 1986, Biomass and production of fishes in natural and channelized streams, *Canadian Journal of Fisheries and Aquatic Sciences*, 43, 1926–34.

31. B. W. Menzel, J. B. Barnum, and L. M. Antosch, 1984, Ecological alterations of Iowa prairie-agricultural streams, *Iowa State Journal of Research*, 59, 5–30.

32. D. R. Hansen, 1971, Stream channelization effects on fishes and bottom fauna in the Little Sioux River, Iowa, in E. Schneberger and J. L. Funk, eds., *Stream channelization: A symposium*, American Fisheries Society Special Publication No. 2, Omaha, Nebr., 29–51.

33. R. Schoof, 1980, Environmental impact of channel modification, *Water Resources Bulletin*, 16, 697–701; C. L. Groen and J. C. Schmulbach, 1978, The sport fishery of the unchannelized and channelized middle Missouri River, *Transactions of the American Fisheries Society*, 107, 412–18; J. C. Congdon, 1971, Fish populations of channelized and unchannelized sections of the Chariton River, Missouri, in E. Schne-

berger and J. L. Funk, eds., *Stream channelization: A symposium,* American Fisheries Society Special Publication No. 2, Omaha, Nebr., 20–28; Gillette, 1972, 890.

34. Committee on Government Operations, 1973, 5, 6–7, 12, 13.

35. Gillette, 1972.

36. Gillette, 1972.

37. R. A. Lohnes, 1997, Stream channel degradation and stabilization: The Iowa experience, in S. S. Y. Wang, E. J. Langendoen, and F. D. Shields, Jr., eds., *Proceedings of the Conference on Management of Landscapes Disturbed by Channel Incision, 1997,* University of Mississippi, Oxford, 35–41; E. H. Grissinger and A. J. Bowie, 1984, Material and site controls of stream bank vegetation, *Transactions of the American Society of Agricultural Engineers,* 27, 1829–35; R. L. Mattingly, E. E. Herricks, and D. M. Johnston, 1993, Channelization and levee construction in Illinois: Review and implications for management, *Environmental Management,* 17, 781–95, 781.

38. Mattingly et al., 1993.

39. R. F. Carline and S. P. Klosiewski, 1985, Responses of fish populations to mitigation structures in two small channelized streams in Ohio, *North American Journal of Fisheries Management,* 5, 1–11; J.-P. Bravard, G. M. Kondolf, and H. Piegay, 1999, Environmental and social effects of channel incision and remedial strategies, in S. E. Darby and A. Simon, eds., *Incised river channels: Processes, forms, engineering and management,* Wiley and Sons, Chichester, U.K., 303–41; F. D. Shields, Jr., S. S. Knight, and C. M. Cooper, 1995, Rehabilitation of watersheds with incising channels, *Water Resources Bulletin,* 31, 971–82; E. A. Dardeau, Jr., and J. C. Fischenich, 1995, Environmental mitigation and the Upper Yazoo projects, *Environmental Geology,* 25, 55–64.

40. D. S. Biedenharn and J. B. Smith, 1997, Design considerations for grade control siting, in S. S. Y. Wang, E. J. Langendoen, and F. D. Shields, Jr., eds., *Proceedings of the Conference on Management of Landscapes Disturbed by Channel Incision, 1997,* University of Mississippi, Oxford, 229–34; M. D. Harvey and C. C. Watson, 1988, Channel response to grade-control structures on Muddy Creek, Mississippi, *Regulated Rivers: Research and Management,* 2, 79–92.

41. R. W. Frazee and D. P. Roseboom, 1997, Development of willow post bank stabilization techniques for incised channelized Illinois streams, in S. S. Y. Wang, E. J. Langendoen, and F. D. Shields, Jr., eds., *Proceedings of the Conference on Management of Landscapes Disturbed by Channel Incision, 1997,* University of Mississippi, Oxford, 313–18; J. C. Burckhard and B. L. Todd, 1998, Riparian forest effect on lateral stream channel migration in the glacial till plains, *Journal of the American Water Resources Association,* 34, 179–84; Shields et al., 1995; Dardeau and Fischenich, 1995.

42. Bravard et al., 1999; P. J. Raven, 1986, Changes of in-channel vegetation following two-stage channel construction of a small rural clay river, *Journal of Applied Ecology,* 23, 333–45.

43. Carline and Klosiewski, 1985; R. A. Kuhnle, C. V. Alonso, and F. D. Shields, Jr., 1997, Geometry of scour holes around spur dikes, an experimental study, in S. S. Y. Wang, E. J. Langendoen, and F. D. Shields, Jr., eds., *Proceedings of the Conference on Management of Landscapes Disturbed by Channel Incision, 1997,* University of Mississippi, Oxford, 283–87; Bravard et al., 1999; Shields et al., 1995; Dardeau and Fischenich, 1995.

44. R. C. Heidinger, 1989, Fishes in the Illinois portion of the upper Des Plaines River, *Transactions of the Illinois Academy of Science,* 82, 85–96.

45. C. M. Cooper, S. Testa, and F. D. Shields, Jr., 1997, Invertebrate response to physical habitat changes resulting from rehabilitation efforts in an incised unstable stream, in S. S. Y. Wang, E. J. Langendoen, and F. D. Shields, Jr., eds., *Proceedings of the*

Conference on Management of Landscapes Disturbed by Channel Incision, 1997, University of Mississippi, Oxford, 887–92; F. D. Shields, Jr., S. S. Knight, and C. M. Cooper, 1998, Rehabilitation of aquatic habitats in warmwater streams damaged by channel incision in Mississippi, *Hydrobiologia,* 382, 63–86.

46. C. C. Watson and D. S. Biedenharn, 1999, Design and effectiveness of grade control structures in incised river channels of northern Mississippi, USA, in S. E. Darby and A. Simon, eds., *Incised river channels: Processes, forms, engineering and management,* Wiley and Sons, Chichester, U.K., 395–422; M. D. Harvey, C. C. Watson, and S. A. Schumm, 1983, Channelized streams: An analog for the effects of urbanization, *1983 International Symposium on Urban Hydrology, Hydraulics and Sediment Control,* University of Kentucky, Lexington, 401–9; M. D. Harvey and C. C. Watson, 1986, Fluvial processes and morphological thresholds in incised channel restoration, *Water Resources Bulletin,* 22, 359–68; C. C. Watson, M. D. Harvey, D. S. Biedenharn, and P. Combs, 1988, Geotechnical and hydraulic stability numbers for channel rehabilitation: Part I, the Approach, in S. R. Abt and J. Gessler, eds., *Hydraulic Engineering,* American Society of Civil Engineers, New York, 120–31.

47. E. H. Grissinger and J. B. Murphey, 1983, Present channel stability and late Quaternary valley deposits in northern Mississippi, *Special Publications of the International Association of Sedimentologists,* 6, 241–50.

48. U.S. Fish and Wildlife Service, 1979, *The Yazoo basin: An environmental overview,* Jackson, Miss.

49. U.S. Fish and Wildlife Service, 1979.

50. U.S. Fish and Wildlife Service, 1979.

51. E. W. Hilgard, 1860, *Report of the geology and agriculture of the State of Mississippi,* Barksdale Printers, Jackson, Miss.; Hilgard quoted in C. C. Watson, N. K. Raphelt, and D. S. Biedenharn, 1997, Historical background of erosion problems in the Yazoo Basin, in S. S. Y. Wang, E. J. Langendoen, and F. D. Shields, Jr., eds., *Proceedings of the Conference on Management of Landscapes Disturbed by Channel Incision, 1997,* University of Mississippi, Oxford, 116; C. R. Thorne, 1999, Bank processes and channel evolution in the incised rivers of north-central Mississippi, in S. E. Darby and A. Simon, eds., *Incised river channels: Processes, forms, engineering and management,* Wiley and Sons, Chichester, U.K., 97–121.

52. J. R. Stilgoe, 1982, *Common landscape of America, 1580 to 1845,* Yale University Press, New Haven, Conn.; E. H. Grissinger, J. B. Murphey, and R. L. Frederking, 1982, Geomorphology of Upper Peters Creek catchment, Panola County, Mississippi: Part II, within channel characteristics, in *Modeling Components of Hydrologic Cycle,* Water Resources Publications, Littleton, Colo., 267–82; E. H. Grissinger and J. B. Murphey, 1983, Morphometric evolution of man-modified channels, in *Proceedings on the Conference on Rivers '83,* American Society of Civil Engineers, New Orleans, La., 273–83.

53. Watson et al., 1997.

54. Thorne, 1999; Watson et al., 1997.

55. S. C. Happ, 1968, Valley sedimentation in north-central Mississippi, in *Proceedings of the Third Mississippi Water Resources Conference,* Jackson, Miss., 1–8.

56. Happ, 1968.

57. F. E. Hudson, 1997, Project formulation of the Demonstration Erosion Control project, in S. S. Y. Wang, E. J. Langendoen, and F. D. Shields, Jr., eds., *Proceedings of the Conference on Management of Landscapes Disturbed by Channel Incision, 1997,* University of Mississippi, Oxford, 120–24.

58. Hudson, 1997.

59. S. R. Abt and C. C. Watson, 1997, Evaluation and evolution of grade control structures, in S. S. Y. Wang, E. J. Langendoen, and F. D. Shields, Jr., eds., *Proceedings of the Conference on Management of Landscapes Disturbed by Channel Incision, 1997,* University of Mississippi, Oxford, 235–40; J. W. Trest, 1997, Design of structures for the Yazoo Basin Demonstration Erosion Control project, in S. S. Y. Wang, E. J. Langendoen, and F. D. Shields, Jr., eds., *Proceedings of the Conference on Management of Landscapes Disturbed by Channel Incision, 1997,* University of Mississippi, Oxford, 1017–22.

60. A. Simon and S. E. Darby, 1997, Disturbance, channel evolution and erosion rates: Hotophia Creek, Mississippi, in S. S. Y. Wang, E. J. Langendoen, and F. D. Shields, Jr., eds., *Proceedings of the Conference on Management of Landscapes Disturbed by Channel Incision, 1997,* University of Mississippi, Oxford, 476–81.

61. Grissinger et al., 1982; A. Simon and R. E. Thomas, 2002, Processes and forms of an unstable alluvial system with resistant, cohesive streambeds, *Earth Surface Processes and Landforms,* 27, 699–718.

62. D. L. Derrick, 1997, Harland Creek bendway weir/willow post bank stabilization demonstration project, in S. S. Y. Wang, E. J. Langendoen, and F. D. Shields, Jr., eds., *Proceedings of the Conference on Management of Landscapes Disturbed by Channel Incision, 1997,* University of Mississippi, Oxford, 351–56.

63. Watson et al., 1997; Hudson, 1997.

64. S. S. Knight, F. D. Shields, Jr., and C. M. Cooper, 1997, Fisheries-based characterization of Demonstration Erosion Control project streams, in S. S. Y. Wang, E. J. Langendoen, and F. D. Shields, Jr., eds., *Proceedings of the Conference on Management of Landscapes Disturbed by Channel Incision, 1997,* University of Mississippi, Oxford, 893–97; Shields et al., 1995.

65. U.S. Fish and Wildlife Service, 1979; D. C. Jackson and J. R. Jackson, 1989, A glimmer of hope for stream fisheries in Mississippi, *Fisheries,* 14, 4–9.

66. U.S. Fish and Wildlife Service, 1979; S. D. Porter, M. A. Harris, and S. J. Kalkhoff, 2001, Influence of natural factors on the quality of Midwestern streams and rivers, *U.S. Geological Survey Water-Resources Investigations Report 00-4288.*

67. U.S. Fish and Wildlife Service, 1979.

68. P. C. Smiley, Jr., C. M. Cooper, K. W. Kallies, and S. S. Knight, 1997, Assessing habitats created by installation of drop pipes, in S. S. Y. Wang, E. J. Langendoen, and F. D. Shields, Jr., eds., *Proceedings of the Conference on Management of Landscapes Disturbed by Channel Incision, 1997,* University of Mississippi, Oxford, 903–8.

69. A. L. Sheldon, 1996, Conservation of stream fishes: Patterns of diversity, rarity, and risk, in F. B. Samson and F. L. Knopf, eds., 1996, *Ecosystem management: Selected readings,* Springer, New York, 16–23.

70. Watson and Biedenharn, 2000; Changnon, 1998.

71. B. Gilbert, 1981, Paddlefish, *Audubon,* 83, 67–71.

72. D. M. Carlson et al., 1982, Low genetic variability in paddlefish populations, *Copeia,* 3, 721–25.

73. D. F. Russell, L. A. Wilkens, and F. Moss, 1999, Use of behavioural stochastic resonance by paddle fish for feeding, *Nature,* 402, 291–94; L. A. Wilkens, D. F. Russell, X. Pei, and C. Gurgens, 1997, The paddlefish rostrum functions as an electrosensory antenna in plankton feeding, *Proceedings of the Royal Society of London,* B 264, 1723–29; G. F. Weisel, 1973, Anatomy and histology of the digestive system of the paddlefish (*Polyodon spathula*), *Journal of Morphology,* 140, 243–56.

74. W. A. Hubert, S. H. Anderson, P. D. Southall, and J. H. Crance, 1984, Habitat suitability index models and instream flow suitability curves: Paddlefish, *U.S. Fish and Wildlife Service, FWS/OBS-82/10.80.*

75. J. H. Crance, 1987, Habitat suitability index curves for paddlefish, developed by the Delphi technique, *North American Journal of Fisheries Management,* 7, 123–30.

76. Crance, 1987; S. J. Zigler, M. R. Dewey, and B. C. Knights, 1999, Diel movement and habitat use by paddlefish in Navigation Pool 8 of the Upper Mississippi River, *North American Journal of Fisheries Management,* 19, 180–87; P. D. Southall and W. A. Hubert, 1984, Habitat use by adult paddlefish in the Upper Mississippi River, *Transactions of the American Fisheries Society,* 113, 125–31.

77. Gilbert, 1981; J. G. Dillard, L. K. Graham, and T. R. Russell, eds., 1986, *The paddlefish: Status, management, and propagation,* American Fisheries Society Special Publication No. 7.

78. K. J. Kilgore, A. C. Miller, and K. C. Conley, 1987, Effects of turbulence on yolk-sac larvae of paddlefish, *Transactions of the American Fisheries Society,* 116, 670–73; C. Gurgens, D. F. Russell, and L. A. Wilkens, 2000, Electrosensory avoidance of metal obstacles by the paddlefish, *Journal of Fish Biology,* 57, 277–90; S. R. Adams, T. M. Keevin, K. J. Kilgore, and J. J. Hoover, 1999, Stranding potential of young fishes subjected to simulated vessel-induced drawdown, *Transactions of the American Fisheries Society,* 128, 1230–34.

79. C. T. Moen, D. L. Scarnecchia, and J. S. Ramsey, 1992, Paddlefish movements and habitat use in Pool 13 of the Upper Mississippi River during abnormally low river stages and discharges, *North American Journal of Fisheries Management,* 12, 744–51.

80. D. T. Gundersen and W. D. Pearson, 1992, Partitioning of PCBs in the muscle and reproductive tissues of paddlefish, *Polyodon spathula,* at the falls of the Ohio River, *Bulletin of Environmental Contamination and Toxicology,* 49, 455–62.

81. J. G. Dillard, L. K. Graham, T. R. Russell, and B. K. Bassett, 1986, Paddlefish—a threatened resource?, *Fisheries,* 11, 18–19; J. R. Waldman, 1995, Sturgeons and paddlefishes: A convergence of biology, politics, and greed, *Fisheries,* 20, 20–21, 49; V. J. Birstein, 1993, Sturgeons and paddlefishes: Threatened fishes in need of conservation, *Conservation Biology,* 7, 773–87; K. Graham, 1997, Contemporary status of the North American paddlefish, *Polyodon spathula, Environmental Biology of Fishes,* 48, 279–89.

82. K. Dabrowski, 1994, Primitive Actinopterigian fishes can synthesize ascorbic acid, *Experientia,* 50, 745–48.

Chapter 6. Trying to Do the Right Thing

1. R. A. Bryson and T. J. Murray, 1977, *Climates of hunger: Mankind and the world's changing weather,* University of Wisconsin Press, Madison; K. P. Nautiyal, 1989, *Protohistoric India,* Agam Kala Prakashan, Delhi, India.

2. N. Smith. 1971, *A history of dams,* Peter Davies, London; E. E. Wohl, 2000, Inland flood hazards, in E. E. Wohl, ed., *Inland flood hazards: Human, riparian and aquatic communities,* Cambridge University Press, Cambridge, U.K., 3–36.

3. V. V. Tepordei, 1987, 1986 aggregate mining data, *Rock Products,* 90, 25–31; M. Sandecki, 1989, Aggregate mining in river systems, *California Geology,* 42, 88–94; G. M. Kondolf, 1994, Geomorphic and environmental effects of instream gravel mining, *Landscape and Urban Planning,* 28, 225–43; G. M. Kondolf, 1997, Hungry water: Effects of dams and gravel mining on river channels, *Environmental Management,* 21, 533–51; M. R. Meador and A. O. Layher, 1998, Instream sand and gravel mining, *Fisheries,* 23, 6–12; D. K. Norman, C. J. Cederholm, and W. S. Lingley, 1998, Flood plains, salmon habitat, and sand and gravel mining, *Washington Geology,* 26, 3–20; L. B. Starnes and D. C. Gasper, 1995, Effects of surface mining on aquatic resources in

North America, *Fisheries,* 20, 20–23; D. O'Neill, 1994, *The firecracker boys,* St. Martin's Griffin, New York; S. Postel, 1999, *Pillar of sand: Can the irrigation miracle last?,* Norton and Company, New York; T. Williams, 2001, Mountain madness, *Audubon,* 103, 36–43.

4. R. LeB. Hooke, 1999, Spatial distribution of human geomorphic activity in the United States: Comparison with rivers, *Earth Surface Processes and Landforms,* 24, 687–92; R. LeB. Hooke, 2000, On the history of humans as geomorphic agents, *Geology,* 28, 843–46.

5. National Academy, 1992, *Restoration of aquatic ecosystems: Science, technology, and public policy,* National Academy Press, Washington, D.C.

6. C. B. Stalnaker, 1994, Evolution of instream flow habitat modeling, in P. Calow and G. E. Petts, eds., *The rivers handbook: Hydrological and ecological principles,* vol. 2, Blackwell Scientific Publications, Oxford, 276–86; B. L. Lamb, 1989, Quantifying instream flows: Matching policy and technology, in L. J. MacDonnell, T. A. Rice, and S. J. Shupe, eds., *Instream flow protection in the West,* Natural Resources Law Center, University of Colorado, Boulder, 23–39; K. D. Bovee, 1986, Development and evaluation of habitat suitability criteria for use in the instream flow incremental methodology, *U.S. Fish and Wildlife Service Biological Report 86 (7),* Instream Flow Information Paper No. 21.

7. T. M. Isenhart, R. C. Schultz, and J. P. Colletti, 1997, Watershed restoration and agricultural practices in the Midwest: Bear Creek of Iowa, in *Watershed restoration: Principles and practices,* American Fisheries Society, Bethesda, Md., 318–34; J. A. Gore and F. L. Bryant, 1988, River and stream restoration, in J. Cairns, Jr., ed., *Rehabilitating damaged ecosystems,* CRC Press, Boca Raton, Fla., 23–38; G. M. Kondolf, 1996, A cross section of stream channel restoration, *Journal of Soil and Water Conservation,* 51, 119–25; J. L. Florsheim and J. F. Mount, 2002, Restoration of floodplain topography by sand-splay complex formation in response to intentional levee breaches, Lower Cosumnes River, California, *Geomorphology,* 44, 67–94; S. J. Bennett, T. Pirim, and B. D. Barkdoll, 2002, Using simulated emergent vegetation to alter stream flow direction within a straight experimental channel, *Geomorphology,* 44, 115–26.

8. N. Gordon, 1995, Summary of technical testimony in the Colorado Water Division 1 trial, *USDA Forest Service General Technical Report RM-GTR-270;* E. E. Wohl, 2001, *Virtual rivers: Lessons from the mountain rivers of the Colorado Front Range,* Yale University Press, New Haven, Conn.; L. G. Witte, 2002, Still no water for the woods, *Rocky Mountain Research Station, Stream Notes,* April 2002.

9. J. R. Luoma, 2001, Blueprint for the future, *Audubon,* 103, 66–69; W. L. Graf, 2001, Damage control: Restoring the physical integrity of America's rivers, *Annals of the Association of American Geographers,* 91, 1–27; R. Showstack, 2002, Missouri River restoration efforts should include adaptive management, report urges, *EOS, Transactions, American Geophysical Union,* 83, 34.

10. M. Grunwald, 2002, Science academy backs restoring Missouri River's natural flow, *Washington Post,* January 10, 2002, A04; American Rivers, 2002, *Newsletter,* spring 2002, vol. 30(2); R. J. Naiman, S. E. Bunn, C. Nilsson, G. E. Petts, G. Pinay, and L. C. Thompson, 2002, Legitimizing fluvial ecosystems as users of water: An overview, *Environmental Management,* 30, 455–67.

11. D. D. Hart, T. E. Johnson, K. L. Bushaw-Newton, R. J. Horwitz, A. T. Bednarek, D. F. Charles, D. A. Kreeger, and D. J. Velinsky, 2002, Dam removal: Challenges and opportunities for ecological research and river restoration, *BioScience,* 52, 669–81; J. Pizzuto, 2002, Effects of dam removal on river form and process, *BioScience,* 52, 683–91; S. Gregory, H. Li, and J. Li, 2002, The conceptual basis for ecological re-

sponses to dam removal, *BioScience*, 52, 713–23; S. L. Rathburn and E. E. Wohl, 2003, Predicting fine sediment dynamics along a pool-riffle mountain channel, *Geomorphology*, 55, 111–24.

12. T. Palmer, 1996, *America by rivers*, Island Press, Washington, D.C.; Graf, 2001.

13. D. M. Thompson, 2002, Long-term effects of instream habitat-improvement structures on channel morphology along the Blackledge and Salmon Rivers, Connecticut, USA, *Environmental Management*, 29, 250–65; D. M. Thompson and G. N. Stull, 2002, The development and historical use of habitat structures in channel restoration in the United States: The grand experiment in fisheries management, *Géographie Physique et Quaternaire*, 56, 45–60.

14. Thompson and Stull, 2002, 49.

15. Thompson and Stull, 2002.

16. G. M. Kondolf, M. W. Smeltzer, and S. F. Railsback, 2001, Design and performance of a channel reconstruction project in a coastal California gravel-bed stream, *Environmental Management*, 28, 761–76; D. M. Thompson, 2002, Channel-bed scour with high versus low deflectors, *ASCE Journal of Hydraulic Engineering*, 128, 640–43; D. M. Thompson, 2003, A geomorphic explanation for channel avulsions following channel relocation in a coarse-bedded river, *Environmental Management*, 31, 385–400.

17. L. B. Leopold, 2001, Let rivers teach us, *Stream Notes July 2001*, Stream Systems Technology Center, USDA Forest Service Rocky Mountain Research Station.

18. V. P. Claassen and M. P. Hogan, 1998, *Generation of water-stable soil aggregates for improved erosion control and revegetation success*, Final Report MR-1242, California Department of Transportation, Sacramento; P. L. Kresan, 1988, The Tucson, Arizona, flood of October 1983: Implications for land management along alluvial river channels, in V. R. Baker, R. C. Kochel, and P. C. Patton, eds., *Flood geomorphology*, John Wiley and Sons, New York, 465–89.

19. Leopold, 2001.

20. Leopold, 2001; Thompson and Stull, 2002.

21. G. M. Kondolf and M. Larson, 1995, Historical channel analysis and its application to riparian and aquatic habitat restoration, *Aquatic Conservation: Marine and Freshwater Ecosystems*, 5, 109–26; Kondolf, 1996.

22. M. Brunke and T. Gonser, 1997, The ecological significance of exchange processes between rivers and groundwater, *Freshwater Biology*, 37, 1–33; J. A. Stanford and J. V. Ward, 1988, The hyporheic habitat of river ecosystems, *Nature*, 335, 64–66; J. A. Stanford and J. V. Ward, 1993, An ecosystem perspective of alluvial rivers: Connectivity and the hyporheic corridor, *Journal of the North American Benthological Society*, 12, 48–60; J. A. Stanford, J. V. Ward, and B. K. Ellis, 1994, Ecology of the alluvial aquifers of the Flathead River, Montana, in J. Gibert, D. L. Danielopol, and J. A. Stanford, eds., *Groundwater ecology*, Academic Press, San Diego, Calif. 367–90; J. V. Ward, 1997, An expansive perspective of riverine landscapes: Pattern and process across scales, *GAIA*, 6, 52–60.

23. R. Newbury, 1995, Rivers and the art of stream restoration, in J. E. Costa, A. J. Miller, K. W. Potter, and P. R. Wilcock, eds., *Natural and anthropogenic influences in fluvial geomorphology*, American Geophysical Union Geophysical Monograph 89, Washington, D.C., 137–49; L. C. De Waal, A. R. G. Large, and P. M. Wade, eds., 1998, *Rehabilitation of rivers: Principles and implementation*, Wiley and Sons, Chichester, U.K.; Kondolf, 1996; K. Prestegaard, 1998, University of Maryland, personal communication; J. Pizzuto, 1998, University of Delaware, personal communication.

24. De Waal et al., 1998.

25. K. B. Mendrop and P. E. Little, 1997, Grade stabilization requirements for incised channels, in S. S. Y. Wang, E. J. Langendoen, and F. D. Shields, eds., *Management of landscapes disturbed by channel incision,* University of Mississippi, Oxford, 223–28; C. C. Watson and D. S. Biedenharn, 1999, Design and effectiveness of grade control structures in incised river channels of north Mississippi, USA, in S. E. Darby and A. Simon, eds., *Incised river channels: Processes, forms, engineering and management,* Wiley and Sons, Chichester, U.K., 395–422.

26. Graf, 2001.

27. Leopold, 2001.

28. H. T. Odum, E. C. Odum, and M. T. Brown, 1998, *Environment and society in Florida,* Lewis Publishers, Boca Raton, Fla.

29. A. G. Warne, L. A. Toth, and W. A. White, 2000, Drainage-basin-scale geomorphic analysis to determine reference conditions for ecologic restoration—Kissimmee River, Florida, *Geological Society of America Bulletin,* 112, 884–99.

30. H. W. Shen, G. Tabios, and J. A. Harder, 1994, Kissimmee River restoration study, *Journal of Water Resources Planning and Management,* 120, 330–49; J. W. Koebel, Jr., 1995, An historical perspective on the Kissimmee River restoration project, *Restoration Ecology,* 3, 149–59; L. A. Toth, D. A. Arrington, M. A. Brady, and D. A. Muszick, 1995, Conceptual evaluation of factors potentially affecting restoration of habitat structure within the channelized Kissimmee River ecosystem, *Restoration Ecology,* 3, 160–80; S. C. Harris, T. H. Martin, and K. W. Cummins, 1995, A model for aquatic invertebrate response to Kissimmee River restoration, *Restoration Ecology,* 3, 181–94.

31. K. Kloor, 2001, Forgotten islands, *Audubon,* 103, 22–27.

32. Odum et al., 1998; M. W. Weller, 1995, Use of two waterbird guilds as evaluation tools for the Kissimmee River restoration, *Restoration Ecology,* 3, 211–24.

33. Odum et al., 1998; T. Levin, 2001, Forever glades: Ebbs and flows of the Great American Wetland, *Audubon,* 103, 38–39, 50–53, 56–57, 60–61.

34. Odum et al., 1998.

35. S. S. Light and J. W. Dineen, 1994, Water control in the Everglades: A historical perspective, in S. M. Davis and J. C. Ogden, eds., *Everglades: The ecosystem and its restoration,* St. Lucie Press, Delray Beach, Fla., 47–84.

36. Levin, 2001, Forever glades; P. Matthiessen, 1990, *Killing Mister Watson,* Random House, New York.

37. Warne et al., 2000; T. Woody, 1993, Grassroots in action: The Sierra Club's role in the campaign to restore the Kissimmee River, *Journal of the North American Benthological Society,* 12, 201–5; T. Levin, 2001, Reviving the river of grass, *Audubon,* 103, 55–61.

38. Warne et al., 2000; Woody, 1993; T. Williams, 2001, Big water blues, *Audubon,* 103, 84–86, 108–11.

39. L. A. Toth, J. T. B. Obeysekera, W. A. Perkins, and M. K. Loftin, 1993, Flow regulation and restoration of Florida's Kissimmee River, *Regulated Rivers: Research and Management,* 8, 155–66; Woody, 1993; F. Graham, Jr., 2001, A wing and a prayer, *Audubon,* 103, 87–91; Koebel, 1995.

40. Woody, 1993; Koebel, 1995.

41. Warne et al., 2000, 885; T. V. Belanger, H. Heck, and C. Kennedy, 1994, Critical dissolved oxygen field studies in the Kissimmee River/floodplain system, *Lake and Reservoir Management,* 9, 55–56.

42. R. Showstack and J. Jacobs, 2003, Funding for Everglades restoration research program "inadequate," says Academy report, *EOS, Transactions, American Geophysical Union,* 84(2), 14; Levin, 2001, Reviving.

43. Toth et al., 1993; W. A. Niering and E. B. Allen, 1995, A cautionary tale, *Restoration Ecology,* 3, 145.

44. Levin, 2001, Reviving.

45. Levin, 2001, Reviving.

46. Levin, 2001, Reviving; M. S. Douglas, 1947, The *Everglades: River of grass,* Rinehart and Co., New York.

47. National Park Service, 2000, *Everglades National Park,* National Park Service brochure, Government Printing Service.

48. Koebel, 1995.

49. M. A. Strawn, 1997, *Alligators: Prehistoric presence in the American landscape,* The Johns Hopkins University Press, Baltimore, Md.

50. F. J. Mazzotti and L. A. Brandt, 1994, Ecology of the American alligator in a seasonally fluctuating environment, in S. M. Davis and J. C. Ogden, eds., *Everglades: The ecosystem and its restoration,* St. Lucie Press, Delray Beach, Fla., 485–505; J. D. Newsom, T. Joanen, and R. J. Howard, 1987, Habitat suitability index models: American alligator, *U.S. Fish and Wildlife Service Biological Report 82 (10.136).*

51. Newsom et al., 1987; G. H. Dalrymple, 1996, Growth of American alligators in the Shark Valley region of Everglades National Park, *Copeia,* 1, 212–15.

52. J. M. Matter, C. S. McMurry, A. B. Anthony, and R. L. Dickerson, 1998, Development and implementation of endocrine biomarkers of exposure and effects in American alligators (*Alligator mississippiensis*), *Chemosphere,* 37, 1905–14.

53. T. Joanen, 1979, *American alligator recovery plan,* Technical Draft, U.S. Fish and Wildlife Service, Washington, D.C.

54. Joanen, 1979.

55. Newsom et al., 1987.

56. Mazzotti and Brandt, 1994; Dalrymple, 1996; C. S. Asa, G. D. London, R. R. Goellner, N. Haskell, G. Roberts, and C. Wilson, 1998, Thermoregulatory behavior of captive American alligators (*Alligator mississippiensis*), *Journal of Herpetology,* 32, 191–97; M. G. Emshwiller and T. T. Gleeson, 1997, Temperature effects on aerobic metabolism and terrestrial locomotion in American alligators, *Journal of Herpetology,* 31, 142–47.

57. D. A. Crain, L. J. Guillette, Jr., A. A. Rooney, and D. B. Pickford, 1997, Alterations in steroidogenesis in alligators (*Alligator mississippiensis*) exposed naturally and experimentally to environmental contaminants, *Environmental Health Perspectives,* 105, 528–33.

58. K. W. Cummins and C. N. Dahm, 1995, Introduction: Restoring the Kissimmee, *Restoration Ecology,* 3, 147–48; J. C. Trexler, 1995, Restoration of the Kissimmee River: A conceptual model of past and present fish communities and its consequences for evaluating restoration success, *Restoration Ecology,* 3, 195–210; C. N. Dahm, K. W. Cummins, H. M. Valett, and R. L. Coleman, 1995, An ecosystem view of the restoration of the Kissimmee River, *Restoration Ecology,* 3, 225–38.

Chapter 7. Thinking in Terms of Rivers

1. A. Ricciardi and J. B. Rasmussen, 1999, Extinction rates of North American freshwater fauna, *Conservation Biology,* 13, 1220–22; Anonymous, 2000, Endangered freshwater species: A conversation with biologist Anthony Ricciardi, *American Rivers,* 28, 6–8; T. Palmer, 1994, *Lifelines: The case for river conservation,* Island Press, Washington, D.C.; *Water: The power, promise, and turmoil of North America's fresh water,* 1993,

National Geographic Special Edition, November; F. Montaigne, 2001, A river dammed, *National Geographic*, 199, 2–33.

2. T. Flannery, 2001, *The eternal frontier: An ecological history of North America and its peoples*, Atlantic Monthly Press, New York.

3. G. H. Schueller, 2001, Eat locally, *Discover*, 22(5), 70–77.

4. S. Steingraber, 1997, *Living downstream: An ecologist looks at cancer and the environment*, Addison-Wesley Publishing Co., Reading, Mass.

5. J. Malbin, 2001, Wasting away, *Audubon*, 103(6), 15.

6. G. P. Marsh, 1864 (1965), *Man and nature: Or, physical geography as modified by human action*, Harvard University Press, Cambridge, Mass.; J. R. Smith, 1950, *Tree crops: A permanent agriculture*, Devin-Adair Company, New York; R. L. Carson, 1962, *Silent spring*, Houghton Mifflin, Boston; D. W. Orr, 1994, *Earth in mind: On education, environment, and the human prospect*, Island Press, Washington, D.C.

7. J. Bourne, 1999, The organic revolution, *Audubon*, 101(2), 64–70; Schueller, 2001.

8. *Action Plan: Component I: Agricultural watershed management*, 2001, USDA Agricultural Research Service, Washington, D.C.

9. H. D. Klein and A. M. Wenner, 2001, *Tiny game hunting*, University of California Press, Berkeley; V. Kratz, 2001, Apples, pears, and pesticides, *Sierra*, 86, 34–35; J. Bourne, 2000, The killer in your yard, *Audubon*, 102(3), 108; Steingraber, 1997, 102.

10. B. Lavendel, 2001, Green house, *Audubon*, 103(2), 72–78.

11. J. Dennis, 1996, *The bird in the waterfall: A natural history of oceans, rivers, and lakes*, Harper Collins Publishers, New York.

12. Steingraber, 1997, 211.

13. L. D. Britsch and J. B. Dunbar, 1993, Land loss rates: Louisiana coastal plain, *Journal of Coastal Research*, 9, 324–38; Committee on Environment and Natural Resources, 2000, *Integrated assessment of hypoxia in the northern Gulf of Mexico*, National Science and Technology Council, Washington, D.C.; D. A. Goolsby, 2000, Mississippi basin nitrogen flux believed to cause Gulf hypoxia, *EOS, Transactions, American Geophysical Union*, 81(29), 321–27; P. N. Spotts, 2001, It could get hotter in Japan thanks to Three Gorges Dam, *The Christian Science Monitor*, April 12, 2001; J. Duyvejonck, 1994, Group warns of upper Mississippi River degradation, *Water Environment and Technology*, 6, 31; R. H. Kesel, 1988, The decline in sediment load of the lower Mississippi River and its influence on adjacent wetlands, *Environmental Geology and Water Sciences*, 11, 271–81; R. H. Kesel, 1989, The role of the Mississippi River in wetland loss in southeastern Louisiana, USA, *Environmental Geology and Water Sciences*, 13, 183–93.

14. Steingraber, 1997, 117.

15. http://water.usgs.gov/nawqa; www.americanrivers.org; www.sierraclub.org/watersentinels.

16. C. E. Hunt and V. Huser, 1988, *Down by the river: The impact of federal water projects and policies on biological diversity*, Island Press, Washington, D.C.

17. B. Doppelt, M. Scurlock, C. Frissell, and J. Karr, 1993, *Entering the watershed: A new approach to save America's river ecosystems*, Island Press, Washington, D.C., xxi.

18. A. Watts, 1975, Tao: The watercourse way, Pantheon Books, New York, 89.

Index